Joachim Nölte

ICP Emission Spectrometry
A Practical Guide

Related titles from Wiley-VCH

José A. C. Broekaert

Analytical Atomic Spectrometry with Flames and Plasmas

2001
ISBN 3-527-30146-1

Helmut Günzler, Alex Williams

Handbook of Analytical Techniques
(2 Bände)

2001
ISBN 3-527-30165-8

Bernhard Welz, Michael Sperling

Atomic Absorption Spectrometry

1998
ISBN 3-527-28571-7

Joachim Nölte

ICP Emission Spectrometry

A Practical Guide

Dr. Joachim Nölte
AnalytikSupport
Simonshofweg 16
88696 Owingen
Germany

Library of Congress Card No. applied for

British Library Cataloguing-in-Publication Data
A catalogue record for this book is available from the British Library.

Bibliografische Information Der Deutschen Bibliothek
Die Deutsche Bibliothek verzeichnet diese Publikation in der Deutschen Nationalbibliografie; detaillierte bibliografische Daten sind im Internet über <http://dnb.ddb. de> abrufbar.

ISBN 978-3-527-30672-5

© 2003 Wiley-VCH Verlag GmbH & Co. KGaA, Weinheim

Printed on acid-free paper.

Composition: J. Nölte

Many thanks to:

Dr. Sabine Mann,
Dr. Peter Taubert
  For their constructive criticism when checking the manuscript
  and for their always helpful cooperation.

The many newcomers and experienced users of ICP emission
  Who have kept me learning by constantly asking questions –
  and who have in the long run encouraged me to write this book.

The manufacturers of ICP emission spectrometers and accessories for the
privilege of using their illustrations and other information material:

- CETAC Technologies
- Glass Expansion
- Jobin Yvon
- PerkinElmer Instruments
- Spectro Analytical Instruments
- Thermo Elemental
- Varian

My family
  For their patience, because a book like this cannot be written during
  "normal" working hours.

Foreword

A long time has passed since Stanley Greenfield and coworkers at Albright & Wilson, and Velmer Fassel and colleagues Richard Wendt, George Dickinson, and Dick Knisely at Iowa State University devised the spectroanalytical inductively coupled plasma in the 1960's. For almost a decade they worked exhaustively for recognition of ICP spectrometry among hesitant manufacturers and potential users. With others Paul Boumans from Eindhoven and Jacques Robin and Jean-Michel Mermet in Lyon joined an enlarging community of researchers generating convincing evidence of the attributes of ICP emission spectroscopy. Fortunately, their efforts were fruitful by the mid-1970's and blossomed in the 1980's. Today ICP emission and mass spectroscopy are mainstay techniques in the analytical chemistry laboratory, field stations, and testing facilities. The numbers of instruments, users, and applications continue to grow, and no competitive spectroscopic technique has proven as truly universal. The ICP source has been a very bright "star" of 20th century analytical chemistry, and the ICP will continue to fulfill this role in the foreseeable decades as the need for ultrasensitive, matrix-free, and ultramicro analyses intensifies.

With tens of thousands of ICP instruments found throughout the world in laboratories for education and training purposes, for routine environmental applications, and for sophisticated isotopic and/or speciation research in human nutrition and environmental forensics, a practical monograph is needed to address instrument basics, design and operation, software features, and applications. Much has been written about ICP spectrometry, and numerous excellent chapters, monographs and texts now compete for a featured place on the ICP library bookshelf. This book provides a unique resource, since it is intended as a tutorial for novices.

Joachim Nölte has extensive practical ICP experience. For more than a decade Dr. Nölte worked as an ICP application chemist at Perkin-Elmer in Überlingen, Germany, where he provided customer support, developed ICP methods, and conducted training courses. Recently, through his own business enterprise, Analytik Support, he offers ICP consulting services and training programs. This experience with early and contemporary ICP emission systems stimulated his desire to write a practice guide that links theory with everyday ICP applications.

This know-how shows up in the eight chapters of this textbook suited for the ICP beginner. Written in a conversation tone and without many equations or detailed theoretical development, the book qualitatively describes the instrument, method development, routine analysis, trouble shooting and maintenance, and brief application examples. General equipment procurement and site preparation considerations also are described succinctly. Distributed through the text are informative side boxes highlighting practical tips for new users, frequently asked questions with reasonable answers, and complementary theory coupled to practice. This material is appropriately illustrated and documented, and it should be a useful resource for anyone interested in carrying out ICP emission spectrometry.

Ramon M. Barnes
University Research Institute for Analytical Chemistry
Amherst, Massachusetts
October 28, 2002

Preface

"Forget about ICP!" – This was the comment written by the completely frustrated author in the laboratory logbook of a newly acquired ICP emission spectrometer in 1981, having come to the end of a long day's work which yielded nothing but a clogged nebulizer, a melted torch, and the discovery that this beautiful Gaussian-shaped peak was really two peaks directly overlapping each other. A large number of samples still waited on the laboratory bench to be analyzed by this wonderful new instrument called ICP, and not a single measurement would reveal their composition. Well, there was actually a second ICP instrument which could have been used instead. This had far better resolution, but unfortunately it drifted so much that it needed to be recalibrated immediately after calibration. The commercially available ICP instruments were in a state one nowadays would consider prototype. Equally, knowledge of how to operate the instruments and understanding of the new technology was minimal. This would have been fine if there had been someone around to ask for help or even if one could have got assistance from a book. Unfortunately, neither existed. However, as a chemist one learns never to give up despite all adversities and tries to learn by the mistakes of a previous try.

The learning curve was composed of acquiring as much theoretical knowledge as possible and performing series after series practical trial and error experiments. As a consequence, the author accumulated a lot of experience, and this was later extended by doing method development and by giving training and advice to users of ICP emission spectrometers in the application laboratory of a leading manufacturer of analytical instruments. Here, the author also spent two years as an internal evaluator during the development of a new instrument, looking after the hardware requirements for all types of applications. Finally, he worked as a freelance consultant for a number of companies, where he acquired a detailed insight into routine laboratories.

Often, the same old problems kept cropping up not only in routine runs but also during method development, and a number of issues seemed unclear to new users of the technique. Consequently, these were the focus of one-week training courses which were given by the author almost every month. During these courses, it appeared that the same questions were being asked, so these became the source of the FAQ (frequently asked questions) list and were integrated into the regular training course. In this book, you will find some selected FAQs in the labeled "boxes". However, the most frequently asked question "Is there is a user-oriented practical guide to ICP-OES?" always had a negative answer. Then, as time went on, the idea of writing the book himself evolved – a book which would be based on the accumulated experience of two decades of experimenting and teaching. So what was at first a vague intention finally materialized in its present form.

Dear reader, I hope that you will find the answers in this book to any questions about ICP OES which may have arisen in your routine laboratory work. And – if after reading it you do not write the sentence which begins this preface in your laboratory journal or whisper it as a secret deep sigh, but instead you get a great many accurate results (and no inaccurate ones) from your ICP emission spectrometer, then your perusal of these pages will have been worth while and the authors efforts will have been well rewarded.

November 2002

Joachim Nölte

Preface

Contents

XII

List of Abbreviations, Acronyms and Symbols

AA	Atomic absorption spectrometry
BEC	Background equivalent concentration
c	Speed of light or concentration
CCD	Charge-coupled device
CID	Charge-injection device
cps	Counts per second
CTD	Charge-transfer device
d	Distance between two grooves on a grating
E	Energy
ETV	Electro thermal vaporization
FIA	Flow injection analysis
FWHH	Full width at half height
h	Planck's constant
HD-PE	High density polyethylene
I	Intensity
ICP	Inductively-coupled plasma
IEC	Inter-element correction
k	Constant
LIPS	Laser induced plasma spectrometry
MHz	Megahertz [million cycles per second]
MS	Mass spectrometry
MSF	Multi-components spectral fitting
n	Optical order
OES	Optical emission spectrometry
PEEK	Polyetheretherketone
PDA	Photodiode array
PFA	Perfluoroalkoxy
PMT	Photomultiplier tube
PP	Polypropylene
ppb	Parts by billion, better: µg/kg or µg/L
ppm	Parts by million, better: mg/kg or mg/L
PTFE	Polytetrafluoroethylene
PVC	Polyvinylchloride
r	Resolution
R	Resolving power
RSD	Relative standard deviation
s	Standard deviation
SCD	Segmented charge-coupled device detector
SRM	Standard reference material
USN	Ultrasonic nebulizer
UV	Ultraviolet
w	Width of a slit
x	Geometric position
Z	Total number of illuminated grooves

XIV

α	Angle of incidence
β	Angle of reflectance
δ	Change
Δ	Difference
λ	Wavelength
ν	Frequency
θ	Blaze angle

1 An Overview

ICP emission spectrometry (ICP-OES) is one of the most important techniques of instrumental elemental analysis. It can be used for the determination of approximately 70 elements in a variety of matrices. Thanks to its versatility and productivity it is used in many different applications, and nowadays it carries the basic workload in many routine laboratories.

This book gives an introduction to the basic principles of ICP emission spectrometry and provides some background information as well as practical hints to the user. This knowledge should enable the reader to appreciate the possibilities and limitations of this analytical technique in order to use it in an optimal way.

Throughout the text, you will find complementary information, which is indicated by a frame around the text. Symbols indicate which type of information is given:

    ✖   Practical tips
    ⓘ   Additional information
    📖   Complementary theory

1.1 Features of ICP-OES

The heart of an ICP emission spectrometer is the plasma, an extremely hot "gas" with a temperature of several thousand Kelvin. It is so hot that atoms and ions are formed from the sample to be analyzed. The very high temperature in the plasma destroys the sample completely, so the analytical result is usually not influenced by the nature of the chemical bond of the element to be determined (absence of chemical interference). In the plasma, atoms and ions are excited to emit electromagnetic radiation (light). The emitted light is spectrally resolved with the aid of diffractive optics, and the emitted quantity of light (its intensity) is measured with a detector. In ICP-OES, the wavelengths are used for the identification of the elements, while the intensities serve for the determination of their concentrations.

Since all elements are excited to emit light in the plasma simultaneously, they can be determined simultaneously or very rapidly one after another. Consequently, the analytical results for a sample can be obtained after a short analysis time. The time needed for a determination depends on the instrument used and is of the order of a few minutes. The fact that all the elemental concentrations are determined in one analytical sequence and not by measuring one series of samples for one element, another series for another element and so on usually makes the technique attractive with respect to speed.

Samples analyzed are normally liquids, occasionally solids and (quite rarely) gases. For the determination of an element, no specific equipment (such as the lamp used in atomic absorption spectrometry) is needed. As a rule, one only needs a calibration solution of the element to be analyzed and a little time for method development. Hence, an existing analytical method can easily be extended to include another element. This makes ICP emission spectrometry very flexible.

ICP-OES has a very large working range, typically up to six orders of magnitude. Depending on the element and the analytical line, concentrations in the range from less than μg/L up to g/L can be determined. Time-consuming dilution steps are therefore rarely needed, which considerably increases the analysis throughput.

Particularly in environmental analysis, the working ranges for many elements correspond to the concentrations normally found in the samples, and this is one of the reasons why this technique is widely used in environmental applications; about half of all users of ICP-OES use it in these or related areas.

Because of the widespread use of this technique in environmental applications, there are a number of standards and regulations that apply. The most important of these are ISO 11885 [1] and EPA Method 200.7 [2]. Moreover, ICP-OES is used in a variety of other applications, such as metallurgy and the elemental analysis of organic substances.

Plasma was first described as an excitation source for atomic spectroscopy in the mid-1960s [3, 4, 5, 6], and the first instrument appeared in research laboratories a decade later. After a further 10 years the technique was commercialized [7, 8, 9]. At first slowly, but then at an increasing rate, ICP emission spectrometers were introduced into routine laboratories. During the same period, the instruments were refined to make them more user friendly [10]. Since the early 1990s, ICP-OES has become the "workhorse" in the modern analytical laboratory [11, 12]. These years also brought a number of significant improvements, most importantly the use of solid-state detectors [13].

1.2 Inductively-Coupled Plasma Optical Emission Spectrometry – the Name Describes the Technique

As a rule, the technique is referred to as ICP or ICP-OES. The latter is the abbreviation for inductively-coupled plasma optical emission spectrometry. The complete name describes or implies the analytic features of this technique: "**Plasma**" describes an ionized gas at very high temperatures. The energy necessary to sustain the plasma is transferred electromagnetically via an induction coil. This method of energy transfer is found in the first part of the name of the technique: "**Inductively-coupled** plasma".

The sample to be analyzed is introduced into this hot gas. As a rule, all chemical bonds are dissociated at the temperature of the plasma, so that the analysis is independent of the chemical composition of the sample. The atoms and ions are excited in the plasma to emit electromagnetic radiation ("light"), which mainly appears in the ultraviolet and visible spectral range. The **emission** of light occurs as discrete lines, which are separated according to their wavelength by diffractive optics, and are utilized for identification and quantification.

Spectrometry is a technique for quantification that uses the emission or absorption of light from a sample. Its goal is the determination of concentrations and differs from qualitative analysis by spectra, which is commonly referred to as spectroscopy [14].

As a rule, in ICP emission spectrometry there is a linear relationship between intensity and concentration over more than 4 to 6 decades. This intensity concentration function depends on a number of parameters, some of which are unknown. Hence, there is a need for empirical proportionality factors. Consequently, in ICP emission spectrometry, these factors have to be determined before the analysis (calibration). One assumes that the slope of the calibration functions does not change between standards and the samples. It is an important prerequisite to ensure good accuracy of the analytical results to prove that this is actually the case. Instrument performances as well as method development have a large influence on this, which can be challenging at times.

Since all atoms and ions emit light simultaneously, ICP-OES is a typical representative of a sample-orientated multi-element technique. This means that the results for the elements in one sample are measured in one step, unlike the element-orientated mode of operation where all samples are examined for one element. After all the samples have been analyzed for the first element, they are then measured in a new series for the next element. A typical representative of the element-orientated mode of operation is classical atomic absorption spectrometry. The advantages of the sample-orientated mode of operation for routine analysis are obvious, since the sample is characterized very quickly.

ⓘ **ICP, ICP-OES, ICP-AES, ICP/AES, ICP emission spectrometry, ICP ES: What is the correct name for this technique?**

The variety of names for this technique reminds one of the tower of Babel. Which version should one follow? We will try to throw some light on this while attempting to trace the origin of the terms used:

Let us start with "ICP", the abbreviation for "inductively-coupled plasma". This is widely accepted. Everyone agrees to use this abbreviation, at least in written communications.

However, the abbreviation "ICP" alone is no longer sufficient as a clear identification of the technique since a similar technique, ICP-MS (ICP mass spectrometry), exists. To distinguish these from each other, it is recommended to add "OES" or "MS" to the abbreviation "ICP" in order to clearly specify which technique is meant. "ICP" should rather be understood as a generic term for both techniques.

The abbreviation "OES" is the short form for "optical emission spectrometry" and has been around for many decades. Originally, it was used in connection with excitation by spark or glow discharge long before inductively-coupled plasma was used analytically.

Since plasma only represents another excitation source, it makes sense to stay with the abbreviation "OES".

Sometimes one finds the name "atomic emission spectrometry" or "AES". Typically, this version is used by users and manufacturers who in many cases have worked with "atomic absorption spectrometry" (AAS) before. The use of the term "atomic" is plausible to some extent since ICP-OES as well as AAS and ICP-MS are categorized under the group name "atomic spectrometry". However, the reference to "atoms" is misleading in a way since most particles in the plasma are ions (Table 2).

Complications could also arise from the fact that the abbreviation "AES" stands for "Auger Electron Spectroscopy". Since this is a completely different technique, it is not likely that these will be confused. However, in order to be sure, is seems wise not to use the abbreviation "AES" for "emission spectrometry".

The light can be diffracted only by optical means. Therefore, it is redundant to include the term "optical". The term "optical emission" is a tautology similar to "cold ice" or "wet water". For this reason ISO 12235 [15] suggests dropping the "O" (and "A") completely. The author follows this logic. This leads to the fact that in addition to the abbreviations "OES" and "AES" there would be yet another form "ES". The likelihood of confusion of these terms increases, so throughout this book "OES" will be used.

In a few rare case, a slash "/" is used to connect "ICP" and "OES". However, to the authors knowledge there is no international norm or regulation that suggests the use of a slash. For the sake of clarity, the use of a slash should therefore cease.

1.3 Distribution of ICP-OES

The first applications of ICP-OES were in metallurgy; however, environmental analysis was the driving force leading to its widespread use in routine laboratories. In addition, the technique is used for a variety of other areas of element determination, as shown in Fig. 1.

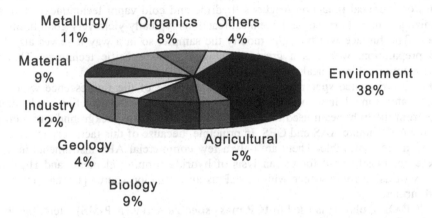

Fig. 1: Use of ICP-OES in different application areas in Germany. Similar patterns can be found in most other countries of the world

With respect to geographical distribution, nearly half of all the instruments in operation worldwide are located in North America. Germany is the next biggest market, with more than 10 %, while Japan and the Netherlands have fewer than 10 % of the installed instruments. Australia, China, Britain and France have shares in the few percent range.

1.4 Related Techniques for Elemental Analysis

Most ICP-OES users gained their initial experience of atomic spectrometry in **atomic absorption spectrometry** (AAS) [16, 17]. AAS is still a technique used in quite a number of laboratories for the basic workload of element determination in a variety of materials. Classical AAS is a single-element technique. AAS uses the absorption of light by atoms which originate from the sample and are in resonance with the light emitted by a specific element lamp. The greater the number of the atoms in the light beam, the higher the absorption, and this is used to measure the concentration. As a rule, the working range is about two orders of magnitude. Flames, furnaces (mainly made of graphite) or chemical reduction reactions (hydride and cold vapor techniques) serve as an atomizing source. Furnaces and chemical reactions typically yield very good limits of detection. The furnace will typically modify the sample, so in a way it serves also for sample preparation, which is a great advantage of this specific technique. This is especially important in clinical applications.

Atomic fluorescence spectrometry (AFS) makes use of the fluorescence which is emitted by atoms in all directions after excitation [18]. Because the detection can be done off-axis from the light beam used to excite the atoms, very low background emission is typical for AFS, unlike AAS and OES. In principle, because of this fact, very low limits of detection are achievable. There are only a few commercial AFS instruments in use, and these are mostly used for the analysis of hydride forming elements and Hg. The possibility of using fluorescence with an ICP as an excitation source [19] has not been pursued further.

As in ICP-OES, plasma is used in **ICP mass spectrometry** (ICP-MS). Here, the ions formed in the plasma are used for quantification. The separation of the ions is carried out (fast) sequentially with a quadrupole (resolution 1 amu) or sector field (resolution of 300 to 10 000 amu with a typical instrument setting of 4 000 amu) or simultaneously using the time-of-flight principle (TOF). The great advantage of ICP-MS consists in its detection power, which is especially high for the high-resolution instruments (if they are operated in a low resolution mode). A greater concentration of dissolved substances in the sample solution may cause clogging of the interface between plasma and high vacuum section of the mass spectrometer. Therefore, the excellent detecting power cannot always be converted to overall better limits of quantification in the sample because in many case the samples must be very highly diluted prior to aspiration into an ICP-MS instrument.

In a **direct current plasma** (DCP) [20], the sample is introduced into a direct current arc. There it is excited to emit light. The region of the arc besides the electric current sustaining the plasma is viewed for quantification. A three-electrode plasma has a cathode and two anodes (Fig. 2). The sample aerosol is introduced via an injector between the two anodes. The analysis with DCP is very susceptible to excitation interference, particularly by easily ionizable elements. In addition, molecular bands will often interfere [21].

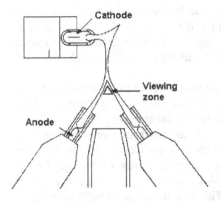

Fig. 2: Schematic representation of a three-electrode direct current plasma (adapted from [22])

A **microwave-induced plasma** (MIP) [23] uses He as the plasma gas. This enables much higher excitation temperatures to be obtained, so that nonmetals are excited particularly well. The MIP is very severely subject to matrix influences, even water. Therefore, it is preferably used for the analysis of gases. The combination with electro-thermal vaporization works very well [24], as does the quantification of hydrides. MIP is an ideal detector for gas chromatography [25, 26, 27].

The plasma temperature of a **capacitively coupled microwave plasma** (CMP) [28] is very low (below 5000 K) [29]. Therefore, excitation interference is quite pronounced. It has been replaced by ICP-OES.

Glow discharge optical emission spectrometry (GDOES) uses the light emitted from a glow discharge formed between a hollow cathode and a sample anode in a reduced-pressure argon atmosphere [30]. Argon cations are formed, which are accelerated in the direction of the negatively charged solid sample. When they hit the surface, atoms from the surface are released and excited. The light from these atoms is used for quantification. GDOES is a technique for surface analysis of electrically conductive materials. Since the composition of the surface has a great influence, a correction must be applied, but this is only successful if all components are known [31].

In **spark optical emission spectrometry** (SOES) [32, 33, 34] part of the material from a metal sample is vaporized with an electric spark. It is further atomized and ionized. The emission during the excitation that takes place is used for quantification. SOES is a fast method to the check the composition of metals [35]. Particularly compact instruments are in use as mobile spectrometers.

Laser-induced plasma spectrometry [(LIPS), also laser-induced breakdown spectroscopy (LIBS)] [36] is a relatively new technique of solid sampling [37]. The irradiation of a solid sample with a laser [38, 39] will cause an immediate transfer of the material into the plasma phase. The radiation emitted from this plasma is used directly for quantification. Since the duration of such an emission signal is very short [40], array spectrometers are typically used in LIPS. Lasers operating in the UV range are preferred since the absorption of lower wavelengths is better than in the visible or infrared region [41]. Both the radiation of the laser to the sample for the generation of the plasma and the radiation emitted can be transferred by fiber optics. Therefore, this technique is particularly attractive for the direct on-line analysis of unapproachable sample material, such as that in nuclear plants [42].

The irradiation of a sample with X-rays triggers the fluorescence of the atoms and ions. Here the lower electron shells (K and L) are excited. The fluorescence emitted is used for qualitative and quantitative analysis in **X-ray fluorescence** (XRF) [43]. In wavelength-dispersive XRF, several dispersive crystals are used to cover the frequency range. In energy-dispersive instruments, the separation is carried out in the detector, which converts the different energies of radiation into electric current or voltage. XRF is used successfully for the analysis of solid samples, especially for large number of samples with small matrix composition changes. The determination of light elements is problematical in XRF. Mutual interference of the elements requires calibration with well matrix-matched standards and advanced processing of intermediate results.

In **total-reflection X-ray fluorescence spectrometry** (TXRF) [44], a liquid sample is placed on a quartz plate and dried. This plate is then placed in the beam of the X ray at a very flat angle so that it is totally reflected. Directly above and normal to this primary beam, a detector [Si (Li)] measures the fluorescence. TXRF is particularly suitable for determining elements with an atomic number of >11 in very low concentrations in small volumes.

1.5 Terms

The element to be determined is called the **analyte**. Its concentration can be in different ranges. If its concentration in the sample is at least 10 %, then it is a **main component**. If its concentration is between 10 % and 0.01 % it is a **minor component**. A component with a lower concentration is called a **trace**. The accuracy and reproducibility normally depend on the magnitude of the concentration to be analyzed. As a rule, only very small deviations are tolerated for main and minor components (typically up to about 1 %), while in trace analysis greater tolerances (up to 10 % and sometimes even more) are accepted.

The tolerances consist of a statistical part, the **reproducibility** of the measurement, which is usually given as standard deviation of the measurement (e.g. intensity in c/s) or in concentration units of the transformed result (e.g. concentration in mg/kg). The reproducibility is frequently also termed the **precision**, although the term is sometimes used in a different context encompassing accuracy and reproducibility. Furthermore, error tolerances include the deviation from the true value, which is referred to as the **accuracy**. It can be determined only on synthetic samples, since true values for real samples cannot be known. However, some well-characterized samples, which were examined by a number of renowned laboratories and evaluated following strict statistical rules, the **standard reference materials** (SRM), exist. The concentration values given in the certificates of these SRMs are generally accepted as "true" values. Quite frequently, the deviations of the measurements from those of the SRMs are used for the determination of the accuracy.

In many cases, the analysis is influenced by the sample components. The sum of all the components of a sample is called the **matrix**. The matrix components may cause an analytical error. This is called **interference**. Since analytical errors are not only linked to the measurement with the ICP instrument itself but are also caused by the complete analytical process, which includes sampling, storage, pretreatment and measurement to name the most important ones, the overall error of the analysis is greater than the measurement tolerances. This concept leads to the idea of the **uncertainty**. The determination of this quantity is quite complex, so that it is very rarely carried out. Obligatory standard methods of performing this calculation that are easily applicable were lacking at the time of printing of this book.

The **sensitivity** is the slope of the calibration function, or expressed mathematically the first derivative of the calibration function [45].

Although not permitted in any standard, the concentration descriptor "parts by million" (ppm) still is quite widespread. Its main disadvantage is the fact that this "unit" does not unambiguously describe whether the concentration refers to particles, mass or volume. Therefore, units as "mg/L" or "mg /kg" are to be preferred.

2 Plasma

A plasma is an ionized gas. It is sometimes referred to as the "fourth" state of aggregation besides the states of aggregation solid, liquid and gas [46]. The states of aggregation can be differentiated by their order at the molecular or atomic level. A phase change is always associated with a qualitative change. As the temperature rises – and particularly at every phase change – the mobility of the particles increases and the order decreases. The plasma therefore exhibits the greatest amount of "disorder" (entropy); an independent motion of electrons and ions characterizes this state.

ⓘ **People refer to the plasma "burning". How can a noble gas like argon burn?**

When argon gas forms a plasma it changes its state of aggregation. This is clearly a physical event and is unlike a chemical reaction in which a gas burns in the presence of oxygen. The "burning" terminology is nevertheless quite widespread – one speaks about a "burning" plasma in a "torch". Although not quite correct, this terminology has become established and its meaning is correctly understood in the context of ICP-OES.

2.1 The Spectrometric Plasma

To sustain the plasma, argon is used in most cases as the operating gas. Positively charged argon ions and negatively charged electrons move independently. The motion of the charged particles (Ar^+ and e^-) follows the acceleration, which is driven by an alternating electromagnetic field. Neutral particles become charged by collisions with the loaded particles and are then accelerated [47].

The transfer of energy in the inductively coupled plasma by the alternating field with an induction coil functions similarly to that in an electric transformer. In this analogy, the primary coil would be the induction coil; the secondary coil contains a current, which corresponds to the core of the plasma itself. Typically, the operating frequency is 27 or 40 MHz. A homogeneous magnetic field is important for the optimal coupling efficiency to the plasma.

The plasma is shaped like a toroid. This form results from the torch geometry, the gas flows and the energy transfer (given by induction coil form, RF power and excitation frequency). This "reversed heart form" is especially suitable to "inject" a gas flow which contains the solution nebulized as an aerosol into the plasma. The carrier gas loaded with the aerosol then forms the so-called analyte channel. The mean dwell time of the sample in the plasma is in order of a few milliseconds.

The temperature of the plasma is not uniform (Fig. 3). It is hottest in the ring-shaped zone inside the induction coil to which the energy of the coil is coupled. Here the plasma reaches temperatures of about 10 000 K. The aerosol-loaded carrier gas is introduced into the center of the plasma ring. From this ring, the aerosol obtains energy. As a result, a temperature gradient exists from the outside to the inside. The carrier gas, which forms the analyte channel, takes the energy and reaches its temperature maximum just after the plasma ring. After the sample has passed through this segment of the plasma, no more energy is supplied to it, so the energy is radiated and the temperature decreases.

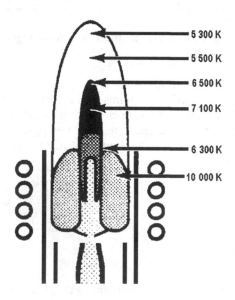

Fig. 3: Temperature distribution in the argon plasma [± 140 K] maintained with an RF power of 1 200 W. (This schematic diagram of the plasma was taken from PerkinElmer Instruments. Temperatures, obtained using Boltzmann plots, were taken from [48].)

✗   **Why do I have to turn on the exhaust before igniting the plasma?**

Nitric oxides and ozone are formed from the nitrogen and oxygen of the air entering into the plasma. Turbulence enhance this effect, which produces small amounts of these harmful gases, which must be removed from the ambient air in order to avoid health hazards affecting laboratory personnel.

2.1.1 The Operating Gas

In commercial ICP instruments, argon is mainly used. As a noble gas with a relatively high atomic number, its electron shell can be easily polarized. Hence, electrons can be more easily released in comparison with noble gases of lower atomic number (Table 1) and molecular gases. In order to ionize molecular gases such as nitrogen, much more energy is necessary. In the early days of ICP, some instruments worked with plasmas using molecular gases such as nitrogen or air. As the cost of the operating gases (e.g. air) decreases, the cost of constructing the RF generator clearly increases. There appears to be no evidence that the analytical performance is better with these systems [49]. Therefore, this concept was not pursued in commercial instruments. Argon is the most common of the "higher" noble gases and thus relative inexpensive.

Table 1: Characteristic data for the noble gases (from [50, 51, 52])

Element	Atomic number	Ionization energy [eV]	Concentration in the atmosphere [ppm]	Typical cost of a 50 L 200 bar bottle [€]
He	2	24.59	5.2	120
Ne	10	21.47	18.2	6 000
Ar	18	15.68	9 340	60
Kr	36	13.93	1.1	20 000
Xe	54	12.08	0.09	200 000

2.1.2 Plasma Torch

The coupling of the energy to the plasma is carried out with an induction coil. Because of the skin effect, the charges are found on the outside in a high-frequency alternating field. As a result, the secondary current, which forms the plasma, extends in the direction of the induction coil. The induction coil is protected by a quartz tube, which shields it from the plasma. Thus, the two electrical circuits (primary in the induction coil, secondary in the plasma core) are separated. However, the tube must be cooled in order to prevent its melting. This is achieved by a strong argon gas flow, which is directed tangentially to the inner surface of the tube. Since other tubes are included in the construction of the plasma torch, this one is referred to as the **outer tube** [54] (Fig. 4). Its diameter is typically 20 mm.

In many cases, this outer tube ends directly above the induction coil. Some torches have an extended outer tube with one or more slits cut into it pointing in the direction of the optics. The advantage of this extension is to keep ambient air away from the plasma [55, 56]. This clearly leads to a lower rate of formation of harmful gases (ozone and nitric oxides). The analytical advantage is that in the absence of nitric oxides, their molecular spectra do not appear. Otherwise, they appear as a strongly structured background in the wavelength region around 230 nm.

✂ **When should I replace the quartz torch?**

If the inner surface of the outer tube becomes rough, then the coolant cannot cool it satisfactorily any longer. In the worst case, the outer tube can melt. It is recommended to replace the torch before this happens because a melted outer tube can also damage the induction coil, causing even greater harm.

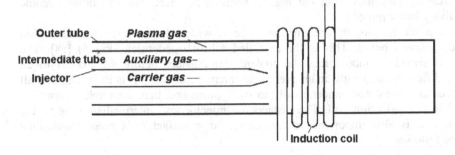

Fig. 4: A schematic drawing of a plasma torch. The torch has an extended outer tube. The short torch variation ends just behind the induction coil (in the figure, to the right) (adapted from a drawing from PerkinElmer Instruments)

The gas for cooling the outer tube is called the **outer gas** or **coolant**. This gas also maintains the plasma, so it is frequently also referred to as **plasma gas**. The coolant or plasma gas flows tangentially into the torch along the outer tube to cool it as efficiently as possible. The plasma gas flows are between 10 and 20 L/min with a typical flow of 15 L/min.

The sample is carried as an aerosol after nebulization in a central tube until just before it enters the plasma. This tube, frequently made from quartz or aluminum oxide, is called the **injector** (tube) or the **inner tube**. The inner gas is also referred to as the **carrier gas** or **nebulizer gas**. Typical carrier gas flows range from 0.6 to 1 L/min, possibly 0.3 to 2 L/min. The speed of the carrier gas flow has an influence on the residence time of the aerosol in the plasma. The longer the dwell time of the sample aerosol in the plasma, the more energy can be absorbed from the plasma, leading to a higher excitation temperature. Consequently, a carrier gas flow as low as possible is preferred. However, there is a limit to reducing the flow. A minimum momentum is necessary for the aerosol to penetrate into the plasma. Therefore, extremely small gas flows cannot be realized. The inside diameter of the injector is typically between 0.8 and 2 mm.

Another tube is located between the two tubes: the **intermediate tube**. This has a typical diameter of 16 mm and serves two purposes: it forces the coolant gas to flow tangentially along the outer tube until shortly before the plasma; it also offers the possibility of introducing another gas flow: the **intermediate** or **auxiliary gas** flow. Its task is to push the plasma away from of the injector tip if necessary. This is important for solutions with a high concentration of dissolved matter (particularly critical just below the solubility limit of the compound). If the injector tip gets too hot, then the solution dries up and solid particles are deposited at this point. In the course of time, the tip of the injector clogs. In addition, if organic solvents are introduced into the plasma, an injector tip that is too hot is unfavorable because the solvents can pyrolyze and form carbon deposits, which also may clog the injector tip. As a rule, the auxiliary gas flow is in the range between 0 and 2 L/min.

At increasing distances from the region where they meet, the gas flows become progressively better mixed [57].

In some torch designs, the intermediate tube is widened shortly before the region where the plasma burns. This is the so-called tulip-shaped torch (Fig. 5) [58]. The widened intermediate tube forces the coolant immediately before the plasma to an accelerated flow which results in an even more intensive cooling. It is important that all three tubes are orientated concentrically to each other and that the torch is installed centrally in the induction coil. The distance of injector and intermediate tube to the induction coil is also important. In any case, the manufacturer's recommendations should be followed.

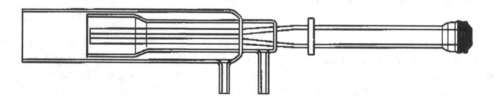

Fig. 5: The drawing shows a tulip-shaped torch. In this torch model, the middle tube is widened immediately before the plasma. The purpose is to accelerate the coolant flow so that it cools the outer tube more efficiently (source: Glass Expansion)

Several years ago, a torch with smaller dimensions and a gas consumption of about 30 % less [59, 60, 61] was used in some instruments but is not pursued anymore.

After prolonged use of the torch, the quartz becomes opaque and the surface becomes rough (devitrification). The amorphous material crystallizes slowly at the high temperatures – just below the melting point. The process is accelerated by alkali metals, which lead to the formation of alkali silicate crystals. Solutions with a high concentration of alkalis therefore inevitably cause the outer tube of the torch to be destroyed relatively quickly. This process is accelerated on an extended outer tube.

> 🛠 **My injector keeps clogging. What could be the reason?**
>
> If you should notice deposits at the injector tip after aspirating solutions with a high dissolved solid content, you should re-adjust the injector so the distance between the induction coil and injector increases. Even fractions of a millimeter can give a considerable improvement. A similar effect can be achieved by increasing the auxiliary gas flow. For dealing with carbon deposits when using organic solvents, proceed similarly.

2.1.3 Ignition of the Plasma

In order to ignite the plasma, the first step is to purge the sample introduction system and the torch with argon to remove molecular gases such as nitrogen or oxygen. These gases absorb so much of the energy supplied by the RF generator that the plasma is destabilized or completely prevented from forming. Next, a high frequency electrical field is applied. This builds up a magnetic alternating field around it. An igniting spark (as a rule a high voltage spark or Tesla spark) produces charge carriers (electrons and argon ions). These charge carriers are then accelerated. The moving ions and electrons form the plasma. Finally, the sample aerosol is introduced into the plasma. Before the analysis can be started, one must wait for the so-called warm-up time to elapse. This time is required to produce a stable signal.

2.2   Excitation to Emit Electromagnetic Radiation (Light)

For the quantitative analysis, the sample aerosol to be examined is introduced into the plasma. Here, the energy from the plasma is transferred to the sample in several steps. At first, the aerosol is dried. The remaining solid then melts and vaporizes. Next, the molecules of the gas formed are dissociated into atoms. The plasma supplies a large amount of energy, so that free electrons from the plasma can remove electrons from the electron shell of an atom by collision, forming positively charged ions. At temperatures of 6 000 to 10 000 K which are typical for an ICP, metals are typically present as ions, while nonmetals and metalloids are only partly ionized. Any surplus energy is used to lift the outer electron of an atom or ion into a higher orbital. This process is called **excitation**.

2.2.1   Emission Lines

The electron stays in the excited state only very briefly: After 10^{-8} s, the electron falls back to an energetically lower orbital. The energy difference between the two energy levels is emitted as electromagnetic radiation predominantly in the ultraviolet wavelength region (190 to 380 nm) and partly in the visible (380 to 800 nm). Since the emission starts from the excited states, there are many "starting points" from where the electrons can fall back. In addition, there are a number of "intermediate points", which clearly increase the number of possibilities for an electron to fall back to a lower state. Spectra in OES are therefore clearly much more line rich than in AAS for example, where the transitions exclusively start from the ground state. Examples of particularly line-rich elements are: Ce 5250 lines, Fe 4400 lines, W 3800 lines, Mo 3400 lines and Cr 3000 lines [62]. The line richness has both advantages and disadvantages. The advantage is the great number of choices of potential analytical lines. The disadvantage is that many lines of the matrix components can cause spectral interference.

In an energy level diagram, the possible energy levels of the orbitals are visualized (Fig. 6). The ordinate indicates the energy difference to the ground state of the atom or ion. The levels of orbitals are indicated by small horizontal lines. The positions of the horizontal lines with respect to the abscissa have no meaning. They are chosen in a way that the possible transitions can be indicated by arrows between the orbital energy levels without intersections of the arrows.

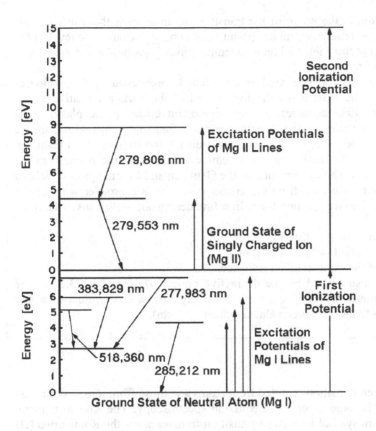

Fig. 6: Energy level diagram of Mg. The energy rises from bottom to top for the atom in the lower part of the diagram. If sufficient energy is supplied to form an ion, a new scale starting with zero is used for the energy levels of the ion in the upper part of the diagram (from [69])

📖 Why is an emission "line" often referred to as a "peak"?

Since the emission results from the transitions between the orbitals, one would expect that the emission image would be extremely narrow because of the discrete nature of the orbitals (a "line"). However, a relatively wide image with clearly visible wings is observed in reality (a "peak"). Even spectra recorded with spectrometers of extremely good resolution have a width of about 1 pm for heavy elements and from 4 to 8 pm for light elements [63]. There are several reasons for this line broadening, and these are partly "natural" and partly a result of the measuring process of the diffractive optics [64, 65].

The natural line width is the width of the transition in an atom without any outside influence. According to Heisenberg, an uncertainty has to be expected at the level of the elementary particles, and this yields a line broadening which is particularly noticeable at the base of the peak (Lorentz profile).

This relatively narrow natural line width is considerably broadened by the motion of the atoms and ions in the plasma. At the high speed of the particles relative to the detector, the Doppler effect becomes the most important effect in the plasma. The relative speed of the particles to the detector causes a frequency shift – and thus a wavelength shift. Since the motion of the particles is distributed statistically, a Gaussian distribution results as the "natural form" of an emission line. The line width is in the range of a few picometers. The combination of the Gaussian and Lorentz profile yields a Voigt profile. Its major difference from the Gaussian profile is a broader wing of the peak at its base [66]. Further reasons for a line broadening are collisions with other particles with

- Free electrons or ions (Stark),
- Neutral particles (van der Waals) or
- Particles of the same kind in resonance.

Line broadening is also caused by the diffractive optics [67] (see Sect. 3.1.1). The optics of the spectrometer typically cause the relatively narrow line widths of a few picometers to broaden to the measured values (typically 10 pm).

To differentiate between transitions emitted from atoms and those from ions, a definition of the transitions has become accepted in emission spectroscopy. The Roman numeral one (**I**) after the element symbol indicates a transition from an atom, the Roman two (**II**) a transition from an ion, and the Roman three (**III**) a transition in an doubly-charged ion. If no entry is made for an element, where the information is given in the same table, it means that the type of transition is unknown.

Sometimes the terms "soft" or "hard" lines are found in the literature. This classification does not correspond to the distinction between transitions of atoms and ions. However, some parallels exist: "hard" lines are exclusively ionic emission lines, while "soft" lines can be either atomic or ionic. "Soft" lines react very strongly to change of the excitation conditions unlike "hard" lines [68].

The classification is based on the norm temperature of 9000 K quoted for the separation of soft and hard lines [69]. A similar distinction can be made based on excitation energies [70].

Spectra of atoms and ions are always line spectra, unlike spectra of molecules, which always have a molecular band structure (many lines in narrow region). Rotational bands are mainly observed (caused by rotation around the bonding axis) in the UV wavelength region.

2.2.2 Energy and Temperature

The energy in the plasma is transferred by collision of an argon ion with another atom. The mean energy of an argon ion is 15.76 eV. The ionization energy of many metals is typically approximately 7–8 eV (Table 2), so there is plenty of energy still available for further excitation of an ion. For this reason, many transitions originating from ions can be found. Depending on the element, there are also a number of transitions from atoms. The alkali metals are ionized very easily, their ionization energy being typically around 4 eV. Metalloids are ionizable a little less readily than most metals, while the nonmetals have typical ionization energies of about 12 eV, which is just below the ionization energy of Ar. Consequently, the latter will not readily form ions in the ICP. The degrees of ionization of selected elements are listed in Table 2.

The amount of emitted electromagnetic radiation is

$$\Delta E = h \cdot v = h \cdot c / \lambda = k / \lambda$$

This means that the emission lines of the visible wavelength region have the lowest energy, which is typical of alkali metals. In the UV range, where most wavelengths used in ICP-OES can be found, the energy differences between the transitions are higher. In the so-called vacuum UV, the transitions with the highest energy can be found. These lines are usually from nonmetals, which are difficult to ionize. Tables 3 and 4 list the transition energies of atomic and ionic emission lines for some selected elements.

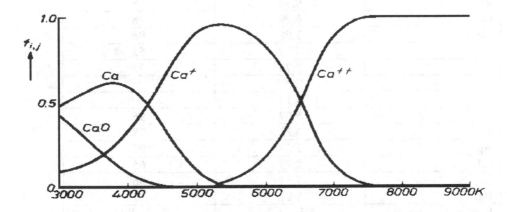

Fig. 7: Influence of temperature on the distribution of species (atoms, ions) (from [69])

Table 2: Ionization energy for selected elements and fraction of the ionized species in the argon plasma (rounded results, from [50])

Element	Ionization energy [eV]	Degree of ionization [%]
Li	5.39	99.9
Be	9.32	75
B	8.29	58
Na	5.14	99.9
Mg	7.65	98
Al	5.99	98
Si	8.15	85
P	10.49	33
S	10.36	14
Cl	12.97	0.9
K	4.34	99.9
Ca	6.11	99
Sc	6.56	99.9
Ti	6.83	99
V	6.75	99
Cr	6.77	98
Mn	7.43	95
Fe	7.90	96
Co	7.88	93
Ni	7.64	91
Cu	7.73	90
Zn	9.39	75
Ga	6.00	98
Ge	7.90	98
As	9.79	52
Se	9.75	33
Br	11.81	5
Rb	4.18	99.9
Sr	5.69	96
Y	6.22	98
Ag	7.58	93
Cd	8.99	65
Sn	7.34	96
Ba	5.21	91
Rare earth elements	5.5 ... 6.2	91 ... 99.9
Au	9.22	51
Hg	10.44	38
Tl	6.11	99.9
Pb	7.42	97

The diagram in Fig. 7 shows the influence of the plasma temperature on the element species, e.g. atom or ion. It illustrates that for relatively low temperatures the number of the free atoms increases with rising temperature, because more and more molecules are dissociated to form atoms. If the temperature is raised further, ions are formed. Hence, the number of the atoms decreases, so that there is a highest number of free atoms at a certain temperature. Since the ions are formed at the expense of the atoms, their concentration increases. Starting at a certain temperature, the number of singly charged ions decreases while the number of doubly charged ions increases. The distribution of the species is reflected in the influence of the plasma temperature on the intensity (Fig. 8). For a given emission line, a temperature can be found where the intensity shows a clear maximum (norm temperature).

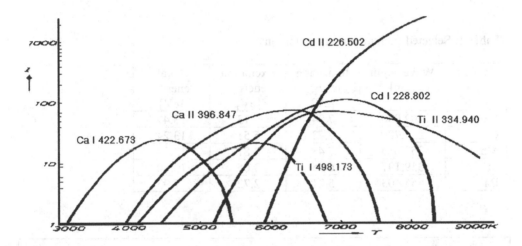

Fig. 8: The intensity distribution of the excitation temperature follows the distribution pattern for the respective species. The temperature at which the maximum intensity is reached is called the norm temperature. Note that the intensity axis is scaled logarithmically (from [69])

Table 3: Selected excitation energies for atoms

Element	Wavelength [nm]	Excitation energy [eV]
S	180.669	6.85
Zn	213.857	5.79
Mg	285.213	4.34
Cu	324.754	3.82
Ca	422.673	2.93
Na	589.598	2.10
K	769.896	1.61

Table 4: Selected excitation energies for ions

Element	Wavelength [nm]	Ionization energy [eV]	Excitation energy [eV]	Total energy [eV]
Mo	202.030	7.34	6.13	13.47
Cu	224.700	7.73	5.51	13.24
Mn	257.610	7.43	4.81	12.24
Zr	339.197	6.95	3.65	10.60
Ba	455.403	5.21	2.72	7.93

📖   **Plasma temperature not a universal value (or how experimental conditions can influence the result)**

The concept of temperature, which reflects the energy content of an object, presupposes a thermodynamic equilibrium. However, this does not exist in the spectrometric plasma [71, 72]. At best, there is a "local thermal equilibrium" (LTE) [73]. In addition, the temperature cannot be measured directly but can be deducted only indirectly [74], for instance by a Boltzmann plot. Depending on whether molecules, atoms, ions or electrons are used as thermometrical species, different values for the temperature are found. Therefore, it makes sense to refer to the gas temperature of the molecules [75], the atomic temperature, the ionic temperature, and the electron temperature [76], which differ from each other at any given location in the plasma [77].

2.2.3 Spectroscopic Properties of the ICP

In order to describe the spectrometric plasma, names for different zones have been defined (Fig. 9) [78, 79]. The **plasma core** is the part to which the energy from the induction coil is coupled. This is the hottest zone of the plasma. The core supplies energy to the remaining parts of the plasma, particularly to the sample in the aerosol, which is introduced via the carrier gas. Sample and carrier gas enter the plasma at approximately room temperature and are heated very rapidly. The first zone of the plasma, where the liquid sample is dried, melted and vaporized, is called **preheating zone**. The next zone is termed the **initial radiation zone**. Here, atoms are formed and excited to the emit light. On further energy uptake, the atoms release electrons to form ions, which also emit light. Ionic transitions predominate in the **normal analytic zone**, which is found outside the plasma core. Since no more energy is supplied in this zone, the temperature drops. The higher the power coupled to the plasma, the closer this zone moves to the induction coil [80]. Finally, the ions recombine with electrons to form atoms, and the atoms react with each other to form molecules. This zone is called the **tail plume**. The entire zone where the sample aerosol is carried through the plasma and which is visible as a brighter color is described as the **analyte channel**. The speed of a particle to move through the plasma is about 20 m/s [81].

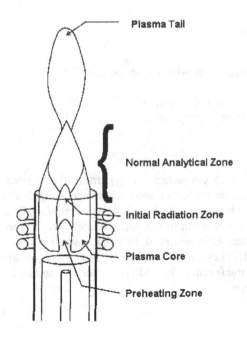

Fig. 9: Spectroscopic zones in the plasma. Typically, the normal analytical zone is viewed radially (from Jobin Yvon)

The emission of light from excited atoms and ions is utilized for the quantification. In a few cases, the light that is emitted in the visible wavelength range can be observed with the eye. Yttrium, for example, emits light in the visible range. The atomic transitions emit red light and the ionic transitions blue light. The distribution of the red and blue light is represented schematically in Fig. 10. Light emitted in the visible range can also be used for diagnostic purposes (see Chapt. 6 "Trouble shooting"). For diagnostics, the intensity distribution of the strong OH molecular bands can also be recorded [82].

When aspirating organic solvents, the initial radiation zone is clearly visible in the form of a green sphere.

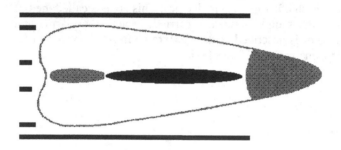

Fig. 10: Schematic representation of a plasma into which a stock solution of yttrium (1 g/L) is introduced:
░░░░░ : Zone of atomic transitions – colored red in the plasma
████ : Zone of ionic transitions – colored blue in the plasma

When matter is carried into the plasma, it consumes energy from the plasma. Consequently, the temperature of the plasma drops in the zone where the sample was introduced. This causes a change in the intensity of the observed emission line. Depending on the temperature of the plasma in relation to the norm temperature, a more or less pronounced decrease in the signal intensity is observed. In a few cases (e.g. alkali metals), a signal enhancement can result. These changes in the sensitivity of the emission signal are a form of **excitation interference**. In addition, interference can be caused by a change in the electron distribution.

ⓘ   **If a concentrated matrix is introduced into the plasma, the intensities change. What is the reason for this? – If I aspirate a highly concentrated acid, the sensitivity changes. Why?**

The dissolved matter in the sample goes through a long process of drying and melting before it is eventually atomized or ionized. These processes consume energy. If the power supplied to the plasma is constant, a drop in the temperature of the plasma results. The more matter introduced, the more pronounced is the observed effect. An increase or decrease in the intensity will be observed, depending on the norm temperature in relation to the plasma temperature.

Absence of excitation interference is indicated by the term **robustness** of the plasma [83]. For quantification of the robustness, the ratio of two magnesium emission lines (atomic and ionic transition) is used. The atomic emission line is at 285.213 nm (background equivalent concentration, BEC = 0.05 mg/L), and the ionic emission line at 280.271 nm (BEC = 0.01 mg/L). The plasma is considered robust if the ratio is 10 or greater. In a robust plasma there is less excitation interference. The Mg I/II ratio depends on a number of parameters like RF power, gas flows such as plasma, auxiliary and particularly nebulizer gas flows, and, in the case of radial viewing, on the viewing height. Figures 11 and 12 show examples of how the robustness of the plasma changes as two of these parameters are altered.

ⓘ   **Plasma robustness and Echelle optics**

In order to quantify the plasma robustness by means of the Mg I/II relation it is presupposed that the optics of the spectrometer do not have any influence on the measured intensities. This presupposition applies to most optical assemblies, but is not true for Echelle optics, which are meanwhile used in many commercial instruments. Here the reflection coefficient changes markedly within a narrow wavelength range of only a few nanometers, so that neighboring lines may have significantly different intensity distributions merely due to the characteristics of the optics. When quantifying the robustness, one should therefore take into account the fact that the characteristics of the Echelle optics can distort the result. In order to get a "correct" estimate of the robustness for Echelle optics, the ratios of the intensities within or between the optical orders must therefore be included as correction factors.

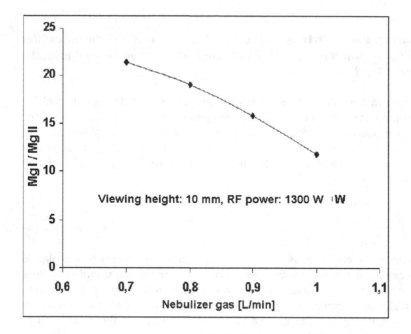

Fig. 11: Influence of the nebulizer gas flow rate on the plasma robustness expressed as a ratio of the intensities of the Mg lines at 280 and 285 nm (generated with a PerkinElmer Optima 4300 DV with axial viewing)

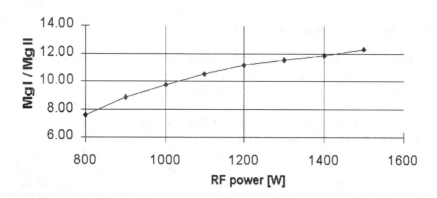

Fig. 12: Influence of the RF power on the plasma robustness expressed as a ratio of the intensities of the Mg lines at 280 and 285 nm (source: Varian)

The plasma temperature results from a number of factors. Mainly these are:

- RF power
- Nebulizer gas flow
- Plasma gas flow
- Auxiliary gas flow
- Sample aspiration rate
- Viewing zone in the plasma.

Therefore, there is a dependence of the intensities of the emission lines on these parameters. Figures 13–17 show this behavior for some analytical wavelengths. To give as much detail as possible, the intensity scales were chosen to accommodate the respective change ranges. When comparing the effects of the various parameters, this has to be taken into account.

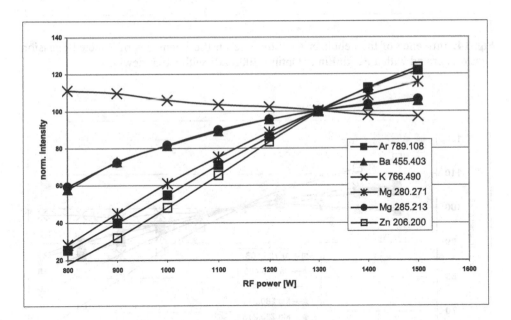

Fig. 13: Influence of the RF power on the intensities of selected emission lines (generated with a PerkinElmer Optima 2000 DV with axial viewing)

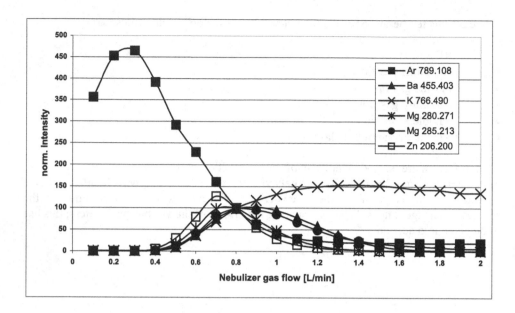

Fig. 14: Influence of the nebulizer gas flow rate on the intensities of selected emission lines (generated with a PerkinElmer Optima 2000 DV with axial viewing)

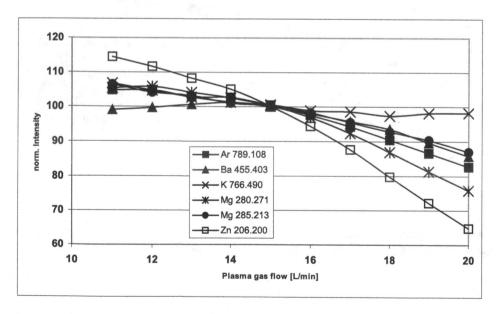

Fig. 15: Influence of the plasma gas flow rate on the intensities of selected emission lines (generated with a PerkinElmer Optima 2000 DV with axial viewing)

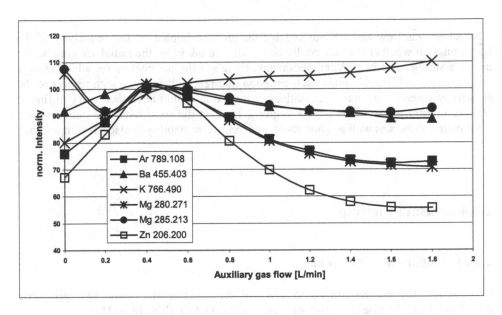

Fig. 16: Influence of the auxiliary gas flow rate on the intensities of selected emission lines (generated with a PerkinElmer Optima 2000 DV with axial viewing)

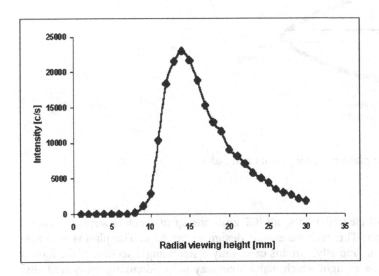

Fig. 17: Influence of the viewing height on the intensity of the Mn emission line at 257.610 nm (generated with a PerkinElmer Optima 2000 DV with axial viewing)

Nebulizer gas flow and RF power have the greatest impact on the sensitivity of the light emission whether viewed axially or radially. In addition, the radial viewing height has in a strong influence. When optimizing, these are the parameters for which change has the largest effect. Although each emission line reacts differently with respect to the sensitivity when the excitation conditions change, one should nevertheless try to find at least compromise conditions for a sub-group. Besides the sensitivity, the independence from matrix effects as well as good short- and long-term stability is also very important.

2.2.4 Plasma Viewing

2.2.4.1 Radial and axial viewing

In principle, the plasma can be viewed (by the spectrometer) from two geometric directions: from the side (radially) or lengthways (axially) (Fig. 18) [84].

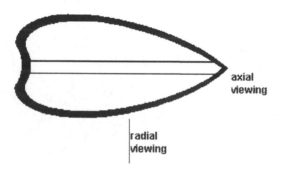

Fig. 18: Directions for plasma viewing: axial and radial

Radial viewing: Until the mid-1990s, all ICP emission spectrometers were constructed with an upright plasma. This was the easiest design technically. The plasma was then viewed from the side or "radially". In this case, only a small angle section of the light is used, namely those rays of light which make their way perpendicularly away (radially) from the analyte channel, which is the symmetry axis of rotation. Of couse, the corresponding section of the plasma is also observed. Instead of "radial viewing", the simplifying but incorrect abbreviation "radial plasma" was quite soon adopted.

The term "side-on viewing" (which happens to be the original term) is also used. In addition to the viewing direction of the traditional upright plasma, one also finds instruments with radial viewing on horizontal plasmas, which sometimes allow both viewing directions. In most instruments, radial viewing allows the user to select a certain viewing height (above the top edge of the induction coil). With this ability, one can restrict the observation zone to a certain segment of the analyte channel, where the excitation conditions are favorable for the selected analytical wavelength. Different zones of the plasma can be viewed (with some instruments even in one analysis run) to ideally reflect the behavior of a certain transition.

Axial viewing: When the plasma is viewed along the analyte channel (which is the symmetry axis of rotation), this is termed "axial viewing". Again, a simplifying and incorrect abbreviation soon started to refer to the axial viewing as "axial" plasma (in analogy to the "radial" plasma). Axial viewing is also called "end-on" plasma.

Axial viewing, which is found only in instruments which have horizontal plasmas [85], generally gives an improvement in the sensitivity by one order of magnitude [86]. The limits of detection are also improved by this factor. The magnitude of the linear working range does not change, but it shifts to lower concentrations [87]. The limits of detection are lowered as the result of a greater number of excited atoms and ions in the viewed section of the plasma and improved signal/background ratio. Even during the pioneer days of ICP, experiments were made with axially viewed plasmas (then referred to as "end-on"). However, the technical and analytical problems prohibited its use for almost two decades [88]. In 1994, the first commercial spectrometers with axial viewing were introduced.

The advantage of improved limits of detection by axial viewing must be set against a number of disadvantages, which blocked the use of the end-on plasma for many years: reduction of the dynamic range and increase in spectral and non-spectral interference [89]. These shortcomings are due to the fact that the plasma tail includes the recombination zone, which acts as an absorption zone for the light. In addition, this zone changes, depending on the matrix. These disadvantages can be compensated for in most cases: the tip of the plasma can be removed with a "shear" gas, by an argon countercurrent or using an interface similar to that used in ICP-MS. By these measures, the observed effects can be much reduced or avoided altogether.

Axial viewing raises not only the sensitivity of the analytes but also that of the matrix components, and the plasma background and its structures, which is illustrated in Fig. 19.

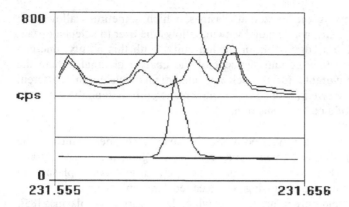

Fig. 19: Spectra illustrating the increase in the intensity of emission signals and background structures on changing from radial viewing (lower spectrum) to axial (upper spectrum). In this figure, the example of the analytical line of Ni at 213.604 nm is shown. The scaling was chosen in order to demonstrate the height of the background and its characteristics for axial and radial viewing

In addition, the axially viewed plasma shows a marked increase in excitation interference compared to radial viewing [90]. This is especially pronounced for easily excitable transitions (e.g. alkali metals) [91] and is known as the alkali effect. This interference affects the entire analyte channel, so that it cannot be controlled in the axially viewed plasma. In this case, radial viewing is preferable.

When viewing the plasma axially, the impairment by background structures is sometimes of greater significance than than the gain in sensitivity, so that the actual gain in terms of detection limits is negligible, and again radial viewing would be preferable. In some spectrometers, both viewing directions are available (examples are shown in Figs. 20 and 62), and this can improve the quality of the results significantly.

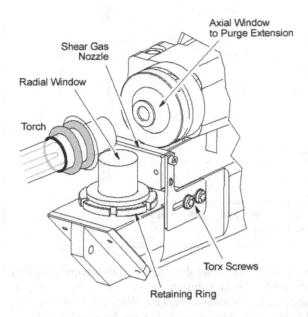

Fig. 20: The plasma is viewed axially by the arranging the transfer optics to be collinear with the analyte channel. The radial view is obtained with the aid of a periscope from the bottom. The shear gas flow (using compressed air) cuts off the tail of the plasma and protects the transfer optics. (This example is taken from the Optima 2000. Source: PerkinElmer Instruments.)

ⓘ   **The alkali effect in the axially viewed plasma**

Alkali metals easily release electrons, thus increasing the electron density in the plasma [92]. If another alkali metal is then introduced into this electron-rich plasma, the free electrons cause a shift of the equilibrium in favor of atoms. As a result, the sensitivity of atomic spectral lines in the plasma increases [93]. This effect is particularly pronounced in the hottest zone of the analyte channel (normal analytical zone) because ionization is less important in the cooler zones of the plasma (namely the initial radiation zone). Because the number of atoms increases, the intensity of atomic transitions increases, so an apparently higher concentration of an alkali metal is found in the presence of high quantities of another alkali metal when viewing the normal analytic zone, as shown in Fig. 21. When viewing the plasma axially, this zone is inevitably included, as when the normal analytic zone of the plasma is viewed radially. If the spectrometer optics are directed to view the initial radiation zone, this effect is not observed, as illustrated in figure 22.

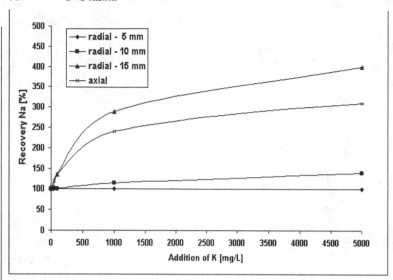

Fig. 21: The alkali effect refers to the observation that the calculated concentration of an alkali metal is impaired in presence of high concentrations of another alkali metal, especially if a zone outside the initial radiation zone is viewed. With axial viewing, the entire area of the analyte channel is included. The alkali effect can be avoided only by optimizing the radial viewing height

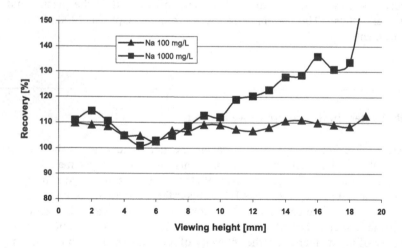

Fig. 22: Influence of the radial viewing height on the recovery of K (1 mg/L) in the presence of Na (100 and 1 000 mg/L). The least influence on the sensitivity is in the lower part of the plasma. In this case, the optimal radial viewing height is found at approx. 5 mm. This point must be determined experimentally by checking the recovery of an alkali metal in presence of different additions of another alkali metal (generated with a PerkinElmer Optima 3200 DV)

2.3 Radio Frequency Generator

The RF (radio frequency) generator supplies the energy for sustaining the plasma. Figure 23 shows a schematic drawing of an RF generator. This produces an electromagnetic high-frequency field in the induction coil with a typical output power of 1000 W. The range of the output power of commercial generators is about 500 to 2000 W. Typically the RF power is set between 800 and 1500 W.

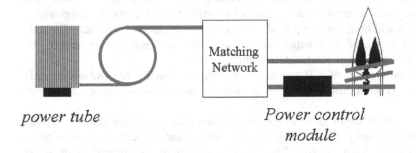

power tube   Power control
                                module

Fig. 23: Schematic diagram of an RF generator (source: Varian)

Since the efficiency of a conventional generator is in the range of 50 %, a considerable amount of waste energy must be removed by cooling with water or air. Blocked air filters for cleaning the ambient air of recirculating coolers substantially lower the cooling efficiency and may result in overheating of the output tube of the RF generator, which dramatically cuts the lifetime of this quite expensive component. The generator is the greatest power consumer of the ICP instrument, with an input power of typically more than 4 KW.

The frequency of the oscillation must be in the MHz range for fundamental reasons. Since very low stray fields cannot always be avoided and malfunctions can occur, only the internationally permitted industry wavebands at 27.12 and 40.68 MHz may be used. After the early days of ICP, a change in the excitation frequency took place in the majority of commercial spectrometers. In the beginning, all instruments were equipped with 27.12 MHz RF generators. Today, more and more RF generators run at a frequency of 40.68 MHz.

At 40 MHz, the secondary current and consequently the plasma core are shifted further to the outside compared to running at 27 MHz. Consequently, the analyte channel becomes wider, which causes a lower speed of transition of the sample particles through the plasma. In addition, the plasma background emission is lowered [94], so the limits of detection are improved [95].

In a frequency-stabilized RF generator, a piezoelectric crystal (quartz resonator) gives very precise impulses for the electric oscillation, while in the non-frequency-stabilized or free-running generator the frequency of the electric oscillation can vary.

As a rule, in frequency-stabilized generators the change in the impedance by entry of matter into the plasma is compensated for by the change in the capacity of large mechanically operated rotary variable capacitors [96]. This technical solution is an electronic control combined with mechanical components. There are many cases where the matching unit reacts too slowly to the matter introduced into the plasma, so that the plasma may become extinguished. Furthermore, the mechanical components wear out over time.

Free-running generators react to the change of the plasma impedance with an altered oscillation frequency without any time delay. Because of this, the generator reacts faster and gives a more robust plasma when samples of strongly changing matrix compositions are introduced. In addition, these types of generators are easier to construct. Thus, the generator can be produced less expensively. Since no matching unit is needed, the construction of the instrument can be more compact.

Instruments in earlier days of ICP typically were frequency-stabilized and occasionally power-stabilized. Today's instruments are mostly frequency free-running and power-stabilized.

The induction coil is exposed to a very severe thermal stress and must be cooled continuously and uniformly. The cooling prevents early wear, since the induction coil is usually manufactured from copper. The cooling is usually performed with water, preferably with a recirculating cooler or in a few cases with plasma gas (argon). A recirculating cooler is preferred since the temperature of the induction coil has a large impact on the energy transfer. A temperature constancy of better than $\pm\,1\,°C$ is desirable.

In order to ensure a constant emission signal, the RF generator power should be stabilized (power-stabilized generator.) One can check this by measuring the stability of argon emission lines. For this test, one measures the reproducibility at typical working conditions but without the influence of the sample introduction system and the carrier gas. The stability typically is 0.1 to 0.3 %.

Generators with power transistors are the latest trend, replacing the thermionic vacuum output tubes. These "solid state" generators are extremely compact (Fig. 24) and have an efficiency of approx. 75 %. It is claimed that better long-term stability should be obtained with them. Unlike thermionic output tubes, solid-state transistors are subject to fundamentally lower wear. A lifetime of 100 000 h is expected from this final stage of the generator, with correspondingly lower operating costs. In addition, solid-state generators do not require any high voltage power supply, unlike thermionic tubes. In contrast, solid-state generators make considerably higher demands on the cooling, since the temperature of the power transistor must remain below 100 °C, unlike the often brightly glowing thermionic tubes of conventional generators. Therefore, very efficient cooling is necessary.

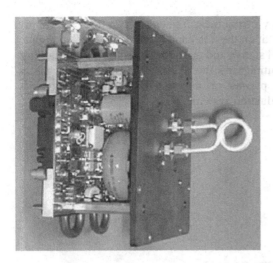

Fig. 24: Solid-state generator with induction coil (source: PerkinElmer Instruments)

As a rule, the control unit for the operating gases (plasma, auxiliary and nebulizer gases) is usually an integral part of the generator. Since the nebulizer gas has the greatest influence on the stability of the emission signal, a mass flow controller is used in a number of instruments to regulate its flow. Especially in recent years, an increasing number of instruments have been constructed in such a way that control of the generator and the gas flows is by the instrument control software.

2.4   Sample Introduction System

The task of the sample introduction system is to modify the sample in such a way that it can be introduced into the plasma without impairing its stability and without influencing the resulting emission signal [97]. This is feasible only for very small particles or droplets.

Normally, liquids are used for analysis by ICP-OES. Therefore, the solution must be converted into small droplets. The solution is sprayed to form a fine aerosol. The smaller the droplet size, the easier it is to dry the droplets and to achieve all the subsequent steps in the plasma. Ideally, the sample solution is sheared into extremely fine droplets during the nebulization process.

If one introduces too large quantities of sample aerosol into the plasma, the latter is extinguished, because the coupled energy is then not sufficient to maintain all processes, from drying to excitation, as described earlier. Amounts of matter which are just below the critical quantity to extinguish the plasma will destabilize it. The plasma then begins to flicker, leading to a poor reproducibility of the measurements. Therefore, the supply of matter to the plasma should be optimized: on the one hand it should be high enough to produce a good sensitivity of the measured signal, and on the other hand the plasma must not be destabilized. The supply of the sample into the plasma in a form which has as small as possible an influence on the plasma or the resulting emission signal is essentially carried out by the sample introduction system.

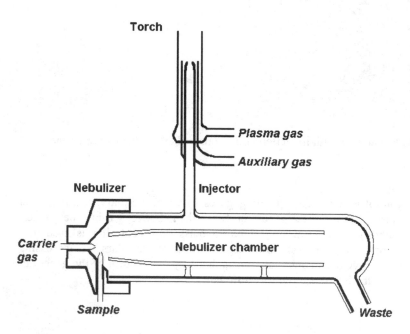

Fig. 25: Schematic diagram of the sample introduction system for a vertical plasma. The sample introduction system contains a Cross-Flow nebulizer and a Scott nebulizer chamber (source: PerkinElmer Instruments)

On careful consideration, one realizes that the physical properties of the sample, such as viscosity, surface tension and density, influence the quantity of the aerosol entering the plasma and consequently the height of the measured signal. Of course, the goal is to minimize the harmful effects of the physical properties of the samples as far as possible.

For all these reasons, the sample introduction has been regarded as of one of the main weak points of this technique since the introduction of ICP-OES [98]. Therefore, different technical solutions for sample introduction systems have been developed to cope with certain difficult applications [99].

The sample introduction system [100] consists of the nebulizer, the nebulizer chamber, and the torch, as shown in the schematic diagram in Fig. 25 and the photograph in

Fig. 26. The pump, typically a peristaltic pump, is sometimes also counted as being part of the sample introduction system.

Fig. 26: Photograph of a sample introduction system equipped with a Cross-Flow nebulizer and a Scott nebulizer chamber. The components are demountable for cleaning or replacement (source: PerkinElmer Instruments)

2.4.1 Nebulizer

The nebulizer converts the sample liquid into an aerosol, which then is transported with a carrier gas flow into the plasma. There are two different basic types of nebulizers commonly used: the **pneumatic nebulizer** and the **ultrasonic nebulizer**. In the pneumatic nebulizer, the carrier gas flow causes the nebulization (so that the carrier gas is also referred to as the nebulizer gas) by forming a negative pressure zone, which breaks up the solution into small droplets. In the ultrasonic nebulizer, the sample solution is pumped onto a small plate vibrating at ultrasonic frequency, where it is thrown off and thus transformed into fine droplets.

2.4.1.1 Pneumatic Nebulizers

There are numerous types and variations of the pneumatic nebulizer. In this book, only the most important ones, which are commonly found in commercial ICP emission spectrometers, will be mentioned: the concentric nebulizer (also Meinhard nebulizer), the Cross-Flow nebulizer, and the Babington-type nebulizer. Pneumatic nebulizers should be operated at least with a carrier gas flow of 0.5 L/min in order to give good analytical performance [101].

The **concentric nebulizer** essentially consists of a capillary for the sample supply, which is concentrically located inside a tube for the gas supply, as illustrated in Fig. 27. The liquid is brought via a tapering capillary to the point where the gas expands in a ring orifice around it. The expanding gas breaks up the liquid and converts it into an aerosol. The concentric nebulizer is self-aspirating because of the Venturi effect.

A self-aspirating nebulizer draws in air if the sample capillary is not placed in a liquid. Too much air in the plasma can destabilize or in the worst case extinguish it. In very rare cases, the torch can melt.

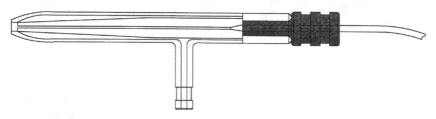

Fig. 27: The sample is supplied to the concentric nebulizer via a tube (from the right). The connection for the nebulizer gas is below (source: Glass Expansion)

The concentric nebulizer constructed by Meinhard was the one supplied with the first commercial instruments. However, the original version of this nebulizer was prone to clogging [102]. A series of design improvements have led to a wide variety of concentric (and other types of "high solids") nebulizers. Members of the series of "original" Meinhard nebulizers are given identifying letters: Type A is the original version. Type C was developed for high concentrations of dissolved matter, differing from type A by having a recessed sample capillary. Type K is like type C, being designed for high dissolved matter contents. The difference is that the surfaces where the nebulization takes place are polished in order not to present any active areas for crystal growth.

It has become customary for manufacturers to give the nebulizers fanciful names, for example the firm Glass Expansion call their concentric nebulizer for high salt contents a "Seaspray" nebulizer. The typical aspiration or sample supply rate of concentric nebulizers is around 1 mL/min.

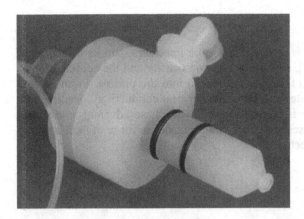

Fig. 28: Concentric nebulizer for low sample consumption. It is metal-free in order to keep the contamination risk low (source: CETAC Technologies)

Besides these types of concentric nebulizers there are special concentric nebulizers optimized for low sample consumption (e.g. 100 µL/min). The most important types are the Micro Concentric Nebulizer (MCN, Fig. 28) and the High Efficiency Nebulizer (HEN) [103]. These nebulizers are used predominantly in ICP-MS. In this technique, the water or the oxygen of the H_2O molecule leads to serious isobaric interference. Therefore, one tries to remove the water by a desolvation as completely as possible. The most efficient removal method is by use of a semipermeable membrane which is swept with a countercurrent flow of dry argon (Fig. 29). Because of the distribution equilibrium, the water is enriched in this gas, while the water content of the sample stream, which is in the form of an aerosol, is depleted. Figure 30 shows such a system.

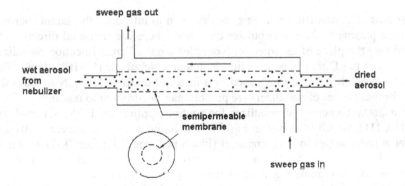

Fig. 29: Schematic diagram of a membrane desolvator (source: CETAC Technologies)

In ICP-OES, the role of the water is different and more important than its role in ICP-MS. It acts rather like a buffer [104], so that the electron density and the cooling effects do not depend simply on the mass of sample introduced into the plasma. Furthermore, the hydrogen generated from the water also causes better heat conductivity, leading to a temperature rise [105]. If the water is completely removed, these advantages are lost. Thus, complete desolvation increases excitation interference, a form of non-spectral interference that is frequently undetected.

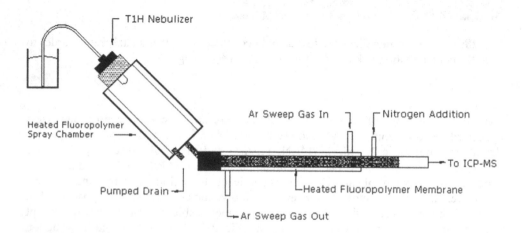

Fig. 30: Nebulizer for low sample consumption with heated nebulizer chamber and membrane desolvation (source: CETAC Technologies)

Special versions of concentric nebulizers are designed to introduce the sample aerosol directly into the plasma without a nebulizer chamber. These are mounted directly into the torch, and take the place of the injector. Examples are the Direct Injection Nebulizer (DIN) [106, 107] or the Direct Injection High-Efficiency Nebulizer (DIHEN) [108]. The sensitivities for some elements are improved by a factor of up to 40 [109]. Nowadays the applications for these types of nebulizers are predominantly confined to research.

The other important pneumatic nebulizer for routine application is the **Cross-Flow nebulizer** [110, 111], which is shown in Fig. 31. The liquid sample is broken up by gas from a jet set at right angles to the sample jet (like a perfume "atomizer"). The relative position of the jets is crucial for the nebulization quality. If the nebulizer is made of oxidizable materials, then oxidizing acids in the sample can attack the jets and cause a poor reproducibility of the measured emission signals [112]. Variations of the Cross-

Flow nebulizer include self-aspirating types, though pump operation is much more common. The typical sample supply rate is also around 1 mL/min.

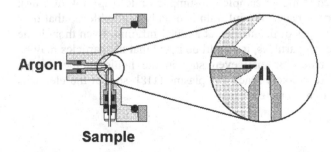

Argon

Sample

Fig. 31: Cross-section of Cross-Flow nebulizer. A detail of the positions of the orifices is shown on the right of the figure (source: PerkinElmer Instruments)

Most of the types of nebulizers mentioned so far tend to clog at higher concentrations of dissolved substances. The exact concentration of dissolved matter depends on the substance and its solubility. Particles floating in the sample solution may also block the sample orifice. A number of nebulizer types were developed specifically for high dissolved solids contents and small particles in the sample solutions. Some fundamental work on these problems was done by Babington, whose name is therefore sometimes used to denote nebulizers that can tolerate these conditions [113].

The **V-groove nebulizer** [114, 115, 116] (Fig. 32) and the **ConeSpray nebulizer** [117] (Fig. 33) are the most important ones for these difficult applications. Both nebulizers have a relatively large aperture for the sample outlet instead of a small jet, and the orifice of the nebulizer gas is continuously rinsed with the sample solution. (Also, this keeps it roughly at room temperature.) Because of these features, sparingly soluble substances rarely precipitate and particles do not block the sample flow.

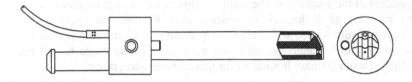

Fig. 32: V-groove nebulizer. Details of the V opening and the exact position of the jets are shown on the right of the figure. The sample enters the V-groove from the top. The nebulizer gas shears the liquid to form an aerosol at the lower jet (source: Glass Expansion)

The two nebulizers differ in the way the sample is fed to the nebulizer gas orifice. In the V-groove nebulizer, the sample runs by gravity down a groove shaped like a V (when viewed from the top) until it passes the nebulizer gas jet. In the ConeSpray nebulizer, a negative pressure of the expanding nebulizer gas draws the sample towards it, thus shearing it into droplets when the liquid meets the expanding gas. Nebulizers of this type are frequently operated at higher sample consumption rates (up to 4 mL/min), although the loss of analytic performance at 1 mL/min is usually negligible, so that they can be operated, just like other types of nebulizers, at about 1 mL/min. Even though the above nebulizers can tolerate small particles, it should be noted that the particles may be removed in the nebulizer chamber (or may even stay in the liquid in the sample container). In any event they are unlikely reach the plasma [118], so that the elements present in these particles may not be detected.

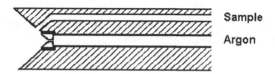

Sample

Argon

Fig. 33: Cross-section of a ConeSpray nebulizer. The sample enters via the upper channel, and the nebulizer gas is fed through the center channel. The opening on the left is rotationally symmetrical and has the appearance of a funnel (source: PerkinElmer Instruments)

Another type of nebulizers uses **high-pressure nebulization** (100–400 bar applied to the sample) [119]. This pressure is achieved with a high-pressure piston pump and a jet of very small diameter. In order to avoid blockages, pre-filtration of the samples is necessary. High-pressure nebulization has an aerosol yield of about 40 %, which is an order of magnitude above that of the common nebulizer types. Because of the high aerosol yield, a subsequent removal of the solvent (desolvation) is required, because otherwise the plasma would be overloaded and become instable.

Apart from the good aerosol yield and the sensitivity gain associated with it, another advantage of the high-pressure nebulization is the fact that the sample supply rate is practically independent of the viscosity of the sample. This aspect is of particular interest for the analysis of oils, as it logical to couple an ICP emission spectrometer incorporating a high-pressure device with HPLC to gain additional information on the type of chemical bonding of the analytes. Although a number of advantages have been published [120], this type of nebulizer has not so far gained wide acceptance.

2.4.1.2   Ultrasonic nebulizer

The **ultrasonic nebulizer** increases the quantity of analyte in the plasma because of the nebulization principle [121]. Typically, the sensitivity rises by a factor of about 10 [122]. Since the reproducibility is comparable to that of pneumatic nebulization, the limits of detection are improved by this factor [123]. In some laboratories, the ultrasonic nebulizer is used routinely. Its main application is in trace analysis of samples which contain very small amounts of dissolved matter [124]. A typical application is the analysis of drinking water [125, 126]. However, analyses of samples of solutions with higher matrix concentration, such as rock samples digested by fusion processes, are possible [127].

A schematic representation of an ultrasonic nebulizer is shown in Fig. 34. The sample is fed to the ultrasonic nebulizer by a peristaltic pump. It then flows down a quartz plate which moves at ultrasonic frequencies (1.4 MHz). The agitation breaks the liquid film into extremely fine droplets, which are thrown into the nebulizer chamber. The carrier gas then takes the droplets to the plasma. Since the nebulization process takes place on the quartz plate, the carrier gas flow can be regulated independently. Figure 35 shows the heart of an ultrasonic nebulizer – the ultrasonic excitation unit (transducer) followed by the nebulizer chamber. The aerosol yield is so high that too much mass would enter the plasma. The solvent (typically water) is therefore partially removed (desolvation).

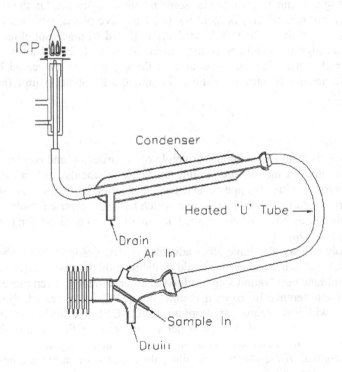

Fig. 34: Schematic diagram of an ultrasonic nebulizer (source: CETAC Technologies)

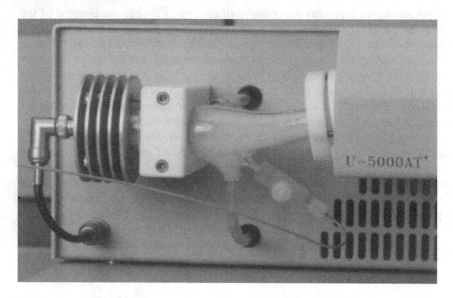

Fig. 35: In this photograph, the heart of an ultrasonic nebulizer is shown. On the left, the ultrasonic excitation unit (transducer) is identifiable by the five plates, which serve for cooling. Adjacent, on the right, is the nebulizer chamber (filled with aerosol cloud). The sample is pumped through the capillary coming from the lower left, and after a bend enters the nebulizer chamber. The further course of the capillary is concealed by the aerosol. The sample aerosol is carried to the right into the desolvation unit (source: CETAC Technologies)

The desolvation step is done following the principle of vaporization and condensation. Modern ultrasonic nebulizers incorporate cooling by Peltier elements, and can thus be put into operation very quickly. It appears that the desolvation step has a substantial influence on the enrichment factor, since a system which has a concentric nebulizer with subsequent desolvation can have an enrichment factor of up to 6, depending on the elements to be analyzed [128].

The removal of solvent may also have other advantages. The desolvation can be used to remove organic solvents which otherwise could destabilize the plasma [129]. For ICP-MS, typically a membrane desolvation step is added in order to remove even more water to suppress isobaric interference by oxygen compounds [130], as described above for concentric nebulizers with low sample consumption. In ICP-OES, the further removal of water will have a disadvantageous effect. The water vapor increases the electron density [131] and maintains constant excitation conditions in the plasma [132]. In contrast, the (almost) complete removal of the solvent will allow the analysis of samples containing very highly volatile solvents [133].

When determining some elements with an ultrasonic nebulizer, it is observed that the sensitivity may depend on the chemical bonding. This was particularly described for As [134]. As(V) is about 30 % more sensitive than As(III). Consequently, the samples should be oxidized with hydrogen peroxide prior to running them with ICP-OES. A similar dependence on the state of oxidation exists for Cr(III) and Cr(VI) [135].

✗   **When aspirating some types of sample, e.g. groundwater, the sensitivity decreases and, sometimes even visibly, the density of the aerosol**

The solution often reveals itself after a closer look at the transducer. One can observe the formation of a coating, which is sometimes resinous. This coating will reduce the amplitude of the vibrating quartz, so that the efficiency of the nebulization is dramatically decreased and less aerosol is generated. Salt deposits and humic acids are typical troublemakers. The remedy for the routine application of samples containing humic acids (such as groundwater) is a complete digestion of the samples. In order to restore the transducer, a suggested solution is to place it flat on fine sandpaper and scrape off the layer by rubbing it on the sandpaper [136].

2.4.2 Nebulizer Chamber

2.4.2.1 Tasks of the Nebulizer Chamber

The nebulizer chamber (also: spray chamber) is part of the sample introduction system for liquids. It is connected to the nebulizer. The prime task of the nebulizer chamber is to remove larger droplets, since these would destabilize the plasma. The aerosol droplets are carried by the nebulizer gas flow. If this flow changes direction, the smallest droplets follow the movement of the gas flow, but larger droplets according the rule of inertia fly on straight ahead and are deposited on the walls of the nebulizer chamber. Depending on the geometry of the nebulizer chamber and the carrier gas speed, only droplets less than a critical diameter ("cut-off diameter") will find their way into the plasma [137].

The residue consisting of the impacted larger droplets accumulates in the lowest part of the nebulizer chamber and is typically removed by a pump [138] or by means of a trap which will allow the liquid to drain away while stopping gases from entering or leaving the sample introduction system. The trap is initially filled with the solvent (as a rule water), which is in turn replaced by the waste. For good sealing, a diameter big enough to build up a hydrostatic pressure which is greater than the pressure of the carrier gas is needed. This passive way of removing the waste requires a steady slope of the tubing to and from the trap. Otherwise, a siphon effect may cause sudden surges associated with pressure changes. These affect the plasma stability and consequently the signal stability. It should be pointed out that the drain-off properties of the nebulizer chamber and the subsequent waste removal have a great influence on the reproducibility of the measurements.

The removal of the larger droplets naturally reduces the efficiency of the sample introduction system. It is claimed that, for a combination of pneumatic nebulizers with conventional nebulizer chambers, around 1 to 2 % of the solution aspirated actually enters the plasma [139]. In order to increase the aerosol yield, impact beads are mounted in some nebulizer chambers. The primary aerosol flow is directed at the impact beads, causing a secondary aerosol that is generated from the impaction. The secondary aerosol has a much finer droplet distribution.

Possible fluctuations in the aerosol flow, which can arise from the peristaltic pump, are reduced in the relatively large volume of the nebulizer chamber. Part of the effect is due to the relatively large volume of the nebulizer chamber, the other being due to the mixing of freshly nebulized sample aerosol with previously generated aerosol circulating in the nebulizer chamber. Not all aerosol leaves the nebulizer chamber in the direction of the injector – some remains in the chamber and circulates back. Consequently, "new" droplets mix with "old" ones in the nebulizer chamber. This causes better reproducibility of the analytical signal, but also causes a longer delay before the signal stabilizes. When assembling the nebulizer and the nebulizer chamber, care should be taken not to prevent free circulation.

The large surface area of the aerosol causes part of the solvent to vaporize in the nebulizer chamber. Therefore, one should take care to keep the temperature of the nebulizer chamber constant in order to achieve good long-term stability. Ideally, the nebulizer chamber should be cooled [140] in order to minimize the quantity of solvent vapor and keep the amount reaching the plasma as low as possible.

✗   Why is it recommended to cool the nebulizer chamber when working with some types of organic solvents?

The energy supplied to the plasma is just sufficient to maintain it in the normal (aqueous) case, when a typical quantity of material is introduced into it. Just to repeat, for conventional nebulization, only up to about 2 % typically reaches the plasma. Assuming that 1 mL/min solution is pumped, then this corresponds to approx. 20 µL/min. Highly volatile organic solvents are, however, almost quantitatively transferred into the gaseous phase, so that a 50-fold amount will enter the plasma. This large quantity of matter would require too much energy, which is not available, and consequently the plasma would be extinguished. Solvents which are not as highly volatile, however, destabilize the plasma. An additional volume flow originating from the liquid evaporated to the vapor phase is added to the nebulizer flow. This very high gas flow literally blows out the plasma.

For applications with highly volatile solvents, there are nebulizer chambers which are enclosed by a cooling jacket, as shown in Fig. 36. A cooled nebulizer chamber reduces the amount of matter that is transferred into the gaseous phase compared to the corresponding amount at room temperature (or even higher temperatures if the nebulizer chamber is inadequately protected against the waste heat of the plasma). Thus, the quantity of matter introduced into the plasma can be limited to an acceptable amount.

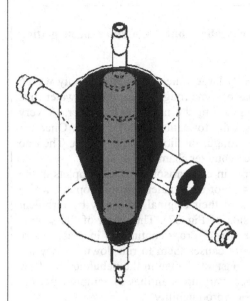

Fig. 36: A cooled cyclonic nebulizer chamber. The cooling liquid is pumped through an outer casing from bottom right to top left (source: Varian)

2.4.2.2 Materials

Materials used to manufacture nebulizer chambers include glass, quartz and polymers (e.g. Ryton or a wide variety of fluorinated polymers). The main properties considered when choosing a material are chemical resistance to the solvents, acids or other reagents used for digestion, and the wettability of the surface of the nebulizer chamber. The latter considerably influences the reproducibility of the analytic signal. Very smooth surfaces cause the solution to form relatively large drops on the surface of the nebulizer chamber. The resulting fluctuations in the aerodynamics of the chamber cause impaired reproducibility.

The large volume of the nebulizer chamber implies a large surface area. This is disadvantageous if the analyte is adsorbed on the material. Depending on the adsorption properties of a given substance on the material, this can lead to extended rinse-out times. The elements B and Hg tend to be strongly adsorbed. Long rinse-out times mean long delays between consecutive measurements. This behavior is also referred to as a **memory-effect**. The sample introduction system in a way acts like a “chromatographic column” (although with a very poor separation performance).

🛠 I have installed a new nebulizer chamber, but have since been getting poorer reproducibility

Who has not observed the formation of relatively large water drops on a freshly waxed car after rainfall? Once the polished surface has become rough again, the rainwater runs off as a thin film. Similar effects also take place in the nebulizer chamber: A very smooth surface of the nebulizer chamber causes the formation of large drops. Once the surface inside the nebulizer chamber becomes rough, the liquid forms a film. Then no droplets or drops can be seen on the wall of the nebulizer chamber.

The visible drops on the surface are very large in comparison with the droplets in the aerosol. The “large” drops partially block the way for the smaller aerosol droplets, which impact on the “large” drops or are thrown out of their original trajectory by turbulent flows and impact on the wall (see left hand side of Fig. 37). Thus, part of the sample aerosol is lost, causing reduced sensitivity. The “large” drops continually increase in size (by coagulation of the smaller droplets). Gravity causes them to run down or they are driven by the nebulizer gas. Consequently, a constant change in the nebulizer gas flow and the aerosol carried by it is observed. Hence, varying quantities of sample enter the plasma, which leads to variable signals and poor reproducibility.

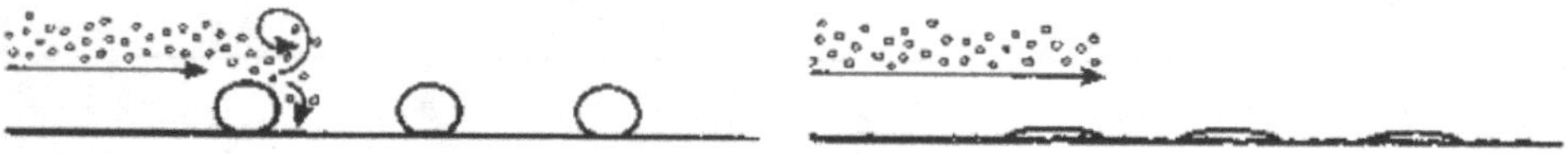

Fig. 37: The formation of comparatively large drops on a very smooth surface of the nebulizer chamber leads to a varying rate of removal of the smallest aerosol droplets (left). A rough surface (right) does not lead to drop formation, so there is no barrier preventing the small droplets from impacting

If droplet formation is noticed on the inside wall of the nebulizer chamber, the surface of the chamber is too smooth and should be roughened by etching (e.g. glass or quartz with approx. 0.5 % hydrofluoric acid). After this procedure, the sample will run off as a liquid film and the aerosol droplets can flow through the nebulizer chamber without barriers, yielding better reproducibly (see right hand side of Fig. 37). As an alternative, it is recommended that a surfactant (e.g. Triton X) be added to all the samples. The surfactant reduces the surface tension and thus prevents the formation of droplets on the wall.

⚒   **Some ICP users add Triton X to all their solutions. What kind of substance is this and what does it do?**

Triton X is a surface-active agent (surfactant). It improves the wettability of the surface by the solution. The surface-active agents normally used in analytical chemistry are highly active. Only few drops of a diluted solution are sufficient to cause the desired effect. Triton X is the best known surfactant used in analytical chemistry. Theoretically, one could use any dishwashing detergent if it were sufficiently pure. Purity is the main reason why most analytical chemists stay with the well-known surfactants such as Triton X. Before using it, its purity should be checked for the planned application

A surfactant can improve reproducibility if the surface of the nebulizer chamber is too smooth and causes droplet formation at the wall. The author recommends treating the cause, in other words etching the surface, rather than constantly treating the symptoms. If the surface is sufficiently rough, no improvement in the reproducibility is observed by the use of a surfactant – in a study on different surfactants, no improvement was found with any of them [141]! This contradicts reports claiming an improvement in reproducibility. In the author's view, this is not a contradiction but only an example of a general statement that apparently unimportant marginal conditions of an experiment can influence the conclusions decisively.

However, there are also applications in which the use of a surfactant is definitely to be recommended. These applications are primarily those where there are different phases in the sample solution (e.g. non-digested blood or slurries). Here it may be necessary to add a surfactant as a phase mediator.

2.4.2.3   Common Types of Nebulizer Chambers

Scott and cyclonic chambers are the most frequently used nebulizer chambers. The **Scott chamber** is a model that has been in use since the early days of ICP. Figure 26 shows a Scott chamber as part of the sample introduction system. It consists essentially of two concentric tubes, of which the outer tube is a little longer and wider than the inner. The total length typically is 15 cm. The outer tube, with a diameter of typically 3 cm, is bordered at one end by a concave curvature. The end cap, which takes the nebulizer, is placed at the other end. The aerosol produced in the nebulizer is collected in the inside tube and transported by the carrier gas to the concave end. Here the gas flow is diverted and the larger droplets are impacted.

The **cyclonic chamber** (Fig. 38) employs the centrifugal forces of a cyclone to separate the aerosol droplets. Not all variations of cyclonic chambers have the ideal form, which includes a baffle tube to guide the aerosol out. In some models, the baffle tube is missing. The nebulizer is inserted at an angle in the cyclonic chamber, so that the aerosol is sprayed tangentially into the chamber (Fig. 39). Use of the cyclonic chamber raises the sensitivity by about 50 % [142], and the limits of detection are reduced by up to a factor of 3 when nebulizers with low sample consumption are used [143].

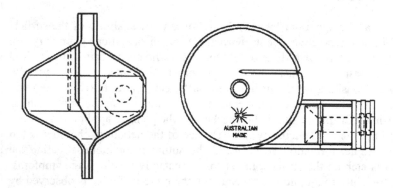

Fig. 38: Schematic diagram of a baffled cyclonic nebulizer chamber. A cross-section is shown on the left of the figure, and a top view on the right (source: Glass Expansion)

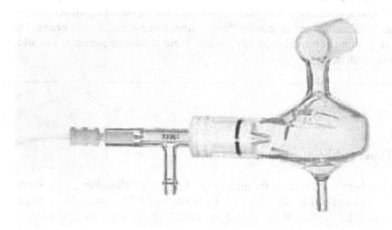

Fig. 39: Cyclonic nebulizer chamber with a concentric nebulizer. The connection at the top right leads to the injector of a horizontal plasma (source: Glass Expansion)

2.4.2.4 Waste from the Nebulizer Chamber

The unused part of the sample must be removed from the nebulizer chamber in such a way that the aerosol-loaded carrier gas is prevented from escaping to the outside and air is prevented from getting into the plasma. Usually this is done passively with a trap filled with liquid or actively by means of a pump. Often, this part of the instrument is not visible, so precautions have to be made to ensure the steady removal of the waste at all times.

In the early days of ICP-OES, passive removal by a trap was the standard way of removing the waste, but in modern systems the pumped waste system is found increasingly often. Both variations are equally good under ideal conditions. However, a siphon effect where the liquid empties suddenly from time to time from a poorly adjusted trap or drain tubes which do not have a constant downward slope will cause an under-pressure, which pulls the plasma down and causes aperiodic signal fluctuations. Hence, the user should take care that the trap is always adjusted correctly to give a good back pressure and the drain tube always has a constant downward slope, as shown in Fig 40.

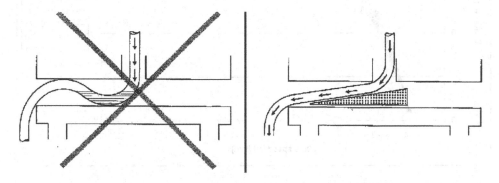

Fig. 40: A constant slope of the waste tube is important to avoid signal fluctuations. Care must be taken to avoid bends

2.4.3 Pump

As a rule, the pump is considered to be part of the sample introduction system. Typically, a peristaltic pump is used. It should move the segments of liquid sample forward as evenly as possible. The more rollers the peristaltic pump has, the higher is the resulting pulsation frequency. Thus, the inevitable pulsations can be averaged in a shorter measurement time or yielding better reproducibility for the same measurement duration.

In addition, it is important that the diameter of the device which presses the pump tube against the rollers is such that the pressure is distributed evenly over the whole of the tube length used for pumping. This pressure should also be optimized, as if it is is too low or too high it can cause poor reproducibility. To get a better tube lifetime, the pressure should be released during times when the instrument is not in us. At low rotational speeds, raising the pump rate causes an increase in the intensity of the analyte signals, as shown in Fig. 41, until it reaches a plateau (at about 1 mL/min for a Cross-Flow nebulizer. At very high speeds, increase in the pump speed causes decreasing sensitivity.

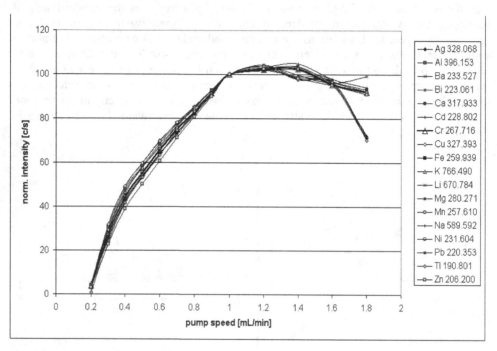

Fig. 41: Influence of the pump speed of a peristaltic pump on the intensities of selected elements (normalized to 100 for a pump rate of 1 mL/min) (data generated with a PerkinElmer Optima 4300 DV equipped with a Cross-Flow nebulizer and a Scott nebulizer chamber)

✂   **How to adjust the pressure on the tube of a peristaltic pump**

There many tips on how this is done. The author has had good results with the following procedure. Release the peristaltic pump pressure completely. With the plasma burning, dip the sample tube into water or into a sample solution and move the tube up and down to catch air segments in the tube. Then slowly increase the pressure until the liquid passes through the tube as constantly as possible. This can be recognized by the captured air segments flowing as evenly as possible.

Some pumps react very strongly to minimal changes in the peristaltic pump pressure. In these cases, a fine adjustment is required by optimizing the reproducibility (RSD) for a calibration solution. Fingertip feel is required here for some pumps, because sometimes barely perceptible twists can have a great influence. When a favorable operating point has been found, it should be marked.

There are also spectrometers where other sources of random variation predominate. Here the influence of the pump is less important, and a fine adjustment by optimizing the reproducibility will not be necessary (or possible).

The quality of the material of the pump tube changes after some operating time. To take care of this (minimal), the procedure described above should be done with a tube which has already been in use for some time.

The choice of the right material for the pump tube is primarily based on its resistance to the solvent used. The manufacturers of tubing normally supply tables of what material is resistant to what solvent. The second important issue is cost, since the pump tube is a consumable item. In spite of the extremely high costs of some materials, these expensive tubes can nevertheless be attractive costwise because their useful life can be much longer, so a break even point will be reached a long time before they become unusable. An example is Viton, which can last for six months, contrasting with a week for other materials (e.g. PVC). Besides purely financial aspects, the added security for routine analysis (especially for overnight runs) also influences the decision in favor of the better materials, since problems due to bad sample transport by the pump will not occur.

✂ Some advice on tubes:

1. For connecting tubes, use Teflon, PEEK or other pure material with preferably low absorbing and exchange behavior.
2. When cutting connecting tubes, cut them at an angle, as it is then easier to insert them into the pump tubes.
3. If you try to insert a connecting tube with a relatively big outside diameter into a pump tube with a relatively small inside diameter, hold the connecting tube with a piece of fine abrasive paper close to the connection to be made. This way, you will have more power in your fingertips and the tube cannot slide away.
4. Sometimes it helps to widen the end of the pump tube with a pipette tip or other conical device, before inserting a connecting tube.

✖   **The concentric nebulizer is self-aspirating. Nevertheless, it is often recommended to operate it with a pump. Why?**

The sample is fed via a very narrow capillary to the jet where it is nebulized. In this capillary, any variations in the viscosity of the solution will have a very strong impact on the sample supply rate [144]. A pump will compensate for this by providing a more reproducible rate of supply of samples or standards regardless of their different viscosities.

✖   **How do I decide that the pump tube should be replaced?**

If replacement is left too late, the tube becomes worn out so badly that it starts to leak and the acid solution trickles onto the pump head. In the worst case, the pump head or even the complete peristaltic pump must be replaced. In order to avoid this scenario, it is recommended to inspect the tube at regular intervals.

A poor pump tube can be spotted by deteriorating reproducibility, but the aim should be to react even before this, so as always to get optimum results with minimum standard deviations.

To achieve this, a regular inspection of the tube is really the only thing to do. The best time for this is at the beginning of a working day. Have a look at the tube! Does it show traces of wear? Are there any deposits visible? If yes, then you should insert a new pump tube. If not, twist the pump tube around his longitudinal axis. The inner diameter of a good tube remains round. A poor pump tube collapses and goes flat, looking a bit like the spiral of a corkscrew.

Sometimes you may forget to release the pump tube at the end of a working day. The tube will then have dents from the rollers imprinted in it. It looks awful, but don't throw it away. Typically, many tubes recover after a couple of days!

In some spectrometers, a fast pump speed can be operated to speed up the analysis. This can be a good idea in order to transport the sample faster to the nebulizer. However, the pump should be put back to its normal rotation speed after the comparatively short time of a few seconds, because the time needed to reach a stable steady-state signal then depends mainly on the time for the changeover in the nebulizer chamber. This time depends on the properties of the sample, the geometry, the material and the surface properties of the nebulizer chamber and the nebulizer gas flow. If the fast speed is reduced only directly before the beginning of the measurement, the reproducibility can deteriorate because of the previous use of the fast pump speed, since the properties of the plasma can change with the change of the quantity of the solution introduced into the plasma. If the fast pump speed is to be used, it is suggested to experimentally determine the time needed for the sample to reach the nebulizer at the fast pump speed and restrict the fast pump time to this duration.

The fast pump speed can be very obstructive when working with highly volatile organic solvents. A sudden burst of a higher supply rate of organic vapors can destabilize the plasma or in the worst case extinguish it.

2.4.4   Other Forms of Sample Introduction

As a rule, in ICP-OES, a liquid sample is aspirated and the measurement is started only after a steady-state signal has been reached. However, there are a number of alternative sample introduction techniques, which are mentioned briefly here. Some of the techniques are used for special samples or applications in routine analysis, while others are employed only in research.

2.4.4.1   Special Techniques for Liquid Samples

Solutions with a very high salt content can clog the nebulizer jets and injector tip. In order to avoid salt deposits, the nebulizer gas can be passed through water (argon humidifier, see Fig. 42).

Fig. 42: The argon humidifier helps to prevent clogging of the nebulizer jets when the sample has a high salt content (source: Thermo Elemental)

Additional information regarding the chemical bond of the analytes can be gained by coupling ICP-OES with chromatographic techniques. Primarily, liquid chromatography (HPLC and ion chromatography) [145, 146, 147, 148, 149, 150] and flow injection techniques are used [106, 151, 152, 153, 154].

A number of sample pretreatment techniques for enrichment or matrix separation have been described [155, 156, 157]. Some of these are carried out "online" [158, 159, 160, 161, 162, 163].

2.4.4.2 Gaseous Samples

Since gaseous substances are introduced almost quantitatively into the plasma, a considerable sensitivity gain can be obtained by converting the analyte into the gaseous phase [164]. "Chemical vaporization" is particularly used for the introduction of hydrides [165] or Hg vapor into the plasma in order to measure them at a much higher sensitivity. In the simplest case, the acidified sample is mixed by means of T-piece with a reducing agent and pumped into the nebulizer [166, 167, 168, 169, 170, 171]. Separation into gaseous and liquid phase then takes place in the nebulizer chamber. Apart from these elements, which can be determined with very much higher sensitivity using this technique, other elements introduced by conventional nebulization can be determined at reduced sensitivity. The usual precautions to ensure a correct result (e.g. pre-reduction) must be taken into account. Another simple variation consists of using part of the nebulizer chamber as a reduction cell, as shown in Fig. 43.

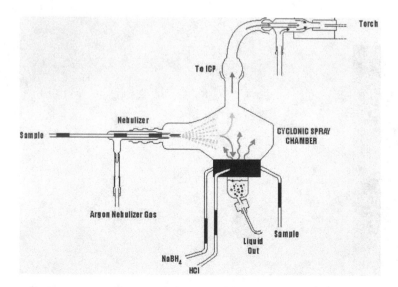

Fig. 43: Use of a modified cyclonic nebulizer chamber for hydride generation and Hg vapor generation (source: Jobin Yvon)

In the literature, a number of devices for hydride generation have been described [172, 173, 174, 175, 176, 177, 178, 179].

When using the amalgamation step, similar detection limits can be reached as in AAS [180].

In addition, other elements, which can form gaseous compounds, e.g. SiF_4, can be determined at a higher sensitivity [181, 182].

2.4.4.3 Solid Sampling

Most samples to be analyzed are originally solids. These must be taken into solution before the analysis. This is frequently associated with a great expense of time and chemicals. Losses can arise or contamination can be carried into the sample solution by the digestion step. Therefore, one would ideally analyze the solids directly.

The simplest way to introduce solids into the plasma is by the slurry technique [183]. Here, a powdered sample is suspended in a solvent and is measured as if it were a solution [184]. This technique is successful only if the particles are particularly small [185, 186, 187, 188, 189] and the bonds of the matrix components are weak [190]. The use of a main component as an internal standard improves the reproducibility by about a factor of 2 [191].

A variation of the slurry technique makes use of a furnace as used in graphite furnace atomic absorption spectrometry. Here the sample is vaporized in the furnace and the vapors are transported to the plasma for analysis [192, 193, 194, 195, 196, 197, 198, 199]. This is known as electrothermal vaporization (ETV) coupling (see Fig. 44).

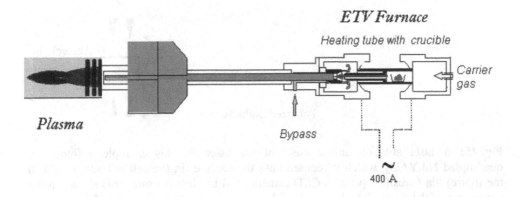

Fig. 44: Coupling of electrothermal vaporization with ICP-OES (source: Thermo Elemental)

Material from a solid sample can be ablated in a finely dispersed form by spark [200] or laser [201] and then introduced into the plasma (laser or spark ablation). While spark ablation was conceived primarily for electrically conductive materials, it can also be used for nonconductive materials [202, 203, 204]. The use of an appropriate internal standard is necessary for quantification [205]. Since there is no solvent in these solid sampling techniques, it cannot act as mediator to stabilize temperature and electron density in the plasma. Hence, excitation interference is an added source of erroneous results [206]. In the solid sampling, a correct analysis requires a calibration with very well matched standards.

Laser ablation can be carried out with any material without special precautions [207]. Figure 45 shows a schematic diagram of a laser ablation unit. The noise level of the ablation process can be used as a measure of the laser energy, which facilitates optimization of the laser conditions. However, the acoustic signal cannot be used as an internal standard signal [208].

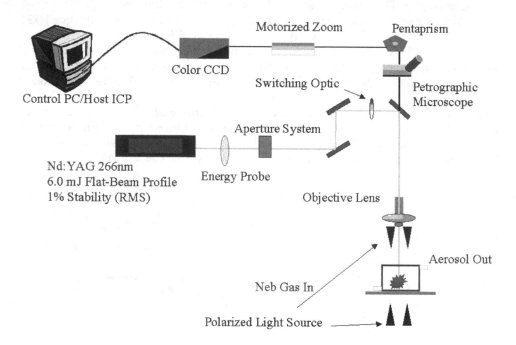

Fig. 45: A laser ablation unit consists of the laser (in this example a frequency quadrupled Nd:YAG), which is focused onto the sample (in the cell at bottom right in the figure) via focusing optics. A CCD camera used to view the sample is shown in the upper part of the figure. The locations of the points to be ablated are stored (on the PC, on the left above). After the start command is given via software, laser shots are directed to these points on the sample. The ablated material is transported by the carrier gas from the sample cell into the plasma (in the figure: "Neb Gas In" and "Aerosol Out") (source: CETAC Technologies)

The laser can be directed onto a defined spot of the sample. Consequently, analytical results for different parts of the sample, i.e. analysis with spatial information, can be generated [209, 210]. The crater produced by the laser shot is very small, so that one almost can describe the laser ablation technique as a "non-destructive" analysis (cf. also Fig. 46). The magnitude and duration of the signal have an impact on the amount of ablated material [211]. The shorter the wavelength of the laser, the lower the influence of the qualities of the sample material [212]. Therefore, lasers that operate with wavelengths in the UV are used preferentially. For UV lasers, the robustness of the plasma has an often underestimated impact on the accuracy of the results.

Fig. 46: Craters formed by laser shots for a repeated analysis for 20 elements [213]. Six craters, which are hardly visible in the upper area of the coin, are sufficient for six independent measurements. Carbon is deposited around the craters in the form of dark borders. These can be easily wiped off

A technique termed "LINA" (Laser-Induced Argon Excitation) uses a laser in order to create a local argon plasma directly over the sample surface [214]. The argon plasma in turn removes a fine aerosol from the sample. The analysis is then performed by ICP. This technique gives the bulk analysis, i.e. the composition of the whole sample without spatial information [215].

3 Optics and Detector of the Spectrometer

The radiation emitted by the atoms and ions of the sample introduced into the plasma are separated spectrally by the optics and the respective emission wavelengths are measured by one or more detectors. Since the emission spectra of most elements are very line rich, spectral interference is the main cause of erroneous measurements. In order to reduce the risk of incorrect measurements, instruments with optics capable of optimal separation of neighboring emission lines are essential in ICP-OES [216].

3.1 Basic Principles of Optics

To characterize an element being analyzed, one wavelength of its emission spectrum, described as the analytical line, is typically used. The spectral separation (dispersion) is usually performed by diffraction gratings. The dispersion process takes place by diffraction of the light at the grating and the interference resulting from this at the focal plane. On diffraction at the grating, enhancements arise at the image plane if the following criterion is valid (**grating equation**):

$$n \, \lambda = d \, (\sin \alpha \pm \sin \beta)$$

where n is the optical order, λ is the wavelength, d is the distance between two grating grooves (grating constant), α is the angle of incidence and β is the angle of diffraction.

Prisms also are used (in addition to gratings) in some optical mounts (Echelle) . The dispersive effect of a prism is based on the different extent of bending of light of different wavelengths on passing from one optical medium (e.g. air) into another (e.g. glass).

3.1.1 Resolution

The selectivity of the analysis depends on the distance of the analyte line from potentially interfering lines. Although the emission signals are commonly referred to as lines, in reality they are not pure lines but have distribution patterns. At a first approximation, these have a Gaussian distribution. In order to distinguish two neighboring wavelengths, there must be a minimum distance between them. Ideally, the base line is reached between two emission lines, but often this is not the case. Using conventional peak processing, it is strongly recommended to use only those lines at which two equally big signals meet at half height. This corresponds to the Rayleigh criterion of spectral resolution.

In order to measure the resolution, the width of a signal is determined at half the net height (cf. Fig. 47). This is frequently known by the abbreviation FWHH (full width at half height) or sometimes FWHM (full width at half maximum). With the help of multivariate processing techniques, it is possible to achieve correct results even in those cases where the classic rules with respect to minimum distance are not followed (see also Sect. 4.2.3.2 "Correction using Multivariate Regression").

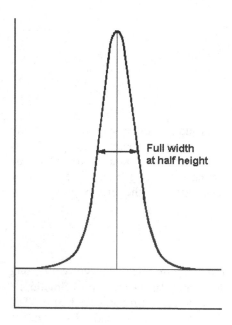

Fig. 47: Idealized emission line where the full width at half height is highlighted. In order to determine the resolution, the width of the line is measured at the height which corresponds to half the net intensity

The spectral resolution is a practical criterion for the spectrometer. The user should preferably determine this or know it already before acquiring a spectrometer. It is important to note that the resolution of some spectrometers (e.g. the Echelle type) can vary over the wavelength range, and therefore a resolution test should be carried out for a number of wavelengths over the full wavelength range. The emission lines have a width of a few picometers (Table 5), so ideally the spectrometer should have this degree of resolution.

Table 5: Line width of selected lines [217]

Element	Wavelength [nm]	FWHH [pm]
Ag	328.068	2.1
Al	396.152	5.1
As	193.696	1.3
Au	208.209	1.3
B	249.773	5.0
Ba	230.527	1.5
Ba	455.403	3.6
Be	313.107	6.2
Ca	393.366	4.1
Cd	228.802	1.5
Co	228.616	5.4
Cr	267.716	2.3
Cu	327.396	7.3
Fe	259.940	2.1
Li	460.286	31.7
Mg	279.553	3.4
Mn	257.610	3.7
Mo	202.030	1.2
Na	589.592	10.3
Ni	231.604	1.9
P	213.618	2.8
Pb	220.353	1.6
Sc	361.384	6.5
Si	251.611	3.0
Sr	421.552	3.1
Ti	334.941	3.1
V	292.402	5.7
W	207.911	1.1
Y	371.030	2.8
Zn	206.200	1.6

It is of vital importance to select a spectrometer with a good resolution to avoid spectral interference. A test typically carried out during the purchase of an instrument particularly refers to the separation of the Cd and As lines at 228.8 nm, as demonstrated as an example in Fig. 48. The narrower the signal, the better the signal/background ratio. This in turn improves the limits of detection.

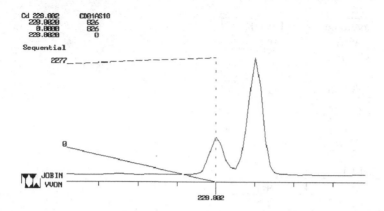

Fig. 48: Test for the optical resolution for the Cd-As interference at 228.8 nm. The two lines are located 10 pm from each other (source: Jobin Yvon)

The resolution *r* is defined by the following equation:

$$r = \text{FWHH} = (\delta\lambda/\delta x)\ (w_1 + w_2)\ /\ 2$$

where w_1 is the width of the entry slit and w_2 is the width of the exit slit.

Good resolution can be achieved by providing
1. A small distance between the grooves of the grating (grating constant)
2. Small entry and exit slits
3. A long focal length
4. High optical order.

ⓘ

In the early days of ICP-OES, the resolution was given in nm, but today it is usually given in pm, reflecting the improved resolution of modern spectrometers.

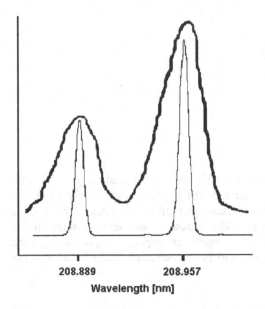

208.889   208.957
Wavelength [nm]

Fig. 49: Improvement in the resolution of ICP emission spectrometers since the commercial introduction of ICP-OES. In this figure, the example of the B doublet around 208.9 nm is shown. The spectra were generated with the PerkinElmer ICP emission spectrometer ICP/5000 (bold line, in the year 1980) and the Optima 3000 (in the year 2000)

The first commercially used ICP spectrometers had a resolution of 30 to 50 pm (Fig. 49). Quite often, spectral interference could not be recognized at this resolution. In order to minimize the risk of erroneous measurements, improvement in the resolution was logically one of the prime goals in the development of new spectrometers. All the four potential improvements listed above were realized in technical developments:

Improvement 1. The best line density (grating constant) can be obtained using holographic gratings. These are produced in photomechanical way. For technical reasons, the line (or groove) density is limited for holographic gratings. Gratings for high-resolution spectrometers have up to 3 600 grooves/mm. The highest line density in a commercial ICP-OES instrument is 4 200 grooves /mm [218].

Improvement 2. Efforts to achieve an improvement in resolution simply by reducing slit widths result in a sometimes drastic sensitivity loss, so that this method is only partially successful. It is important that the slit width should be matched to the resolving power of the optics. If the slit width is too large, not only is the resolution impaired but also the signal/background ratio decreases. Consequently, the limits of detection also decrease, as illustrated in Fig. 50.

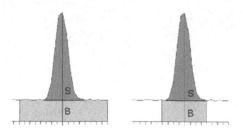

Fig. 50: The detector receives emitted light not only due to the analyte peak but also from the plasma background radiation. If the exit slit is disproportionately large, the analyte signal (S) is recorded along with a sizeable amount of the background (B), as shown on the left. This leads to a poor signal/background ratio and consequently to poorer limits of detection. The right hand figure shows an improved signal/background ratio due to a narrower exit slit, which will yield better detection limits (source: Jobin Yvon)

Improvement 3. Even before the analytical use of an inductively coupled plasma, emission spectrometers were equipped with optics with very long focal lengths (sometimes several meters). Some of these instruments were quite bulky, and the geometric location of the wavelengths on the detector(s) used to react very sensitively to temperature changes. If the processing was done with a photographic plate this was not a major disadvantage, because the complete spectrum was taken and all lines showed the same wavelength shift. On the other hand, a photomultiplier system only registers the intensity at a single point. In these systems, if a wavelength drift or shift occurs, the peak maximum may not fall completely on the detector, resulting in lower and irreproducible readings [219]. Therefore, in these cases efforts must be made to ensure that the wavelength positions do not change.

Improvement 4. In addition to the measures described, the use of higher diffraction orders has also been used to improve resolution. In the case of holographic gratings, this is associated with light loss. Furthermore, an increase in other types of spectral interference is observed because of the overlap between different optical orders. Consequently, interfering orders must be removed. As a rule, this is done with filters. A successful way of using higher orders can be found in Echelle optical mounts. In this design, a mechanically ruled grating, which has a preferred direction of reflection due to the precisely produced correct groove angle, is used for the higher order diffraction, which is then associated with good light throughput and minimum order overlap.

  Why isn't the measured resolution as good as the theoretically calculated resolution?

The actual measured resolution of a spectrometer can be worse than the theoretically calculated resolution, usually because of improper focusing. An image (emission line) which is out of focus appears to be wider than it should be from calculation. In addition, rotation of a solid-state detector will cause the light to fall on neighboring pixels, and a worse apparent separation results between two neighboring pixels. Furthermore, a smoothing function, which aims at improving the reproducibility [220] and the appearance of the spectrum, ultimately has a detrimental effect on the resolution achieved by the optics.

3.1.2   Relevant Optical Terms

The resolution is based on the design of the optics. To give a better understanding, we digress somewhat at this point to discuss basic optics.

The **resolving power** ($R = \lambda_1 / \Delta\lambda$) is expressed as the minimum wavelength difference $\Delta\lambda$ at a wavelength λ_1 which is necessary to separate the slit image at λ_1 from that at $\lambda_1 + \Delta\lambda$. Furthermore, the resolving power is given by the equation

$$R = n \cdot Z$$

where Z is the number of illuminated grooves.

The **reciprocal linear dispersion** $\delta\lambda/\delta x$ gives the change of the wavelength λ per amount of change of geometric distance x in the image plane (at the detector). The geometric slit width is the effective mechanical width [in mm] of the entry and exit slits.

An **optical grating** is a mirror, which has grooves (sometimes confusingly also referred to as lines) with equidistant spacing (grating constant), as illustrated in Fig. 51. The diffraction of the light takes place at the grooves. Here the light is reflected in all directions. The light is radiated in form of waves. Interferences or reinforcements take place. Reinforcements appear where wave crests meet.

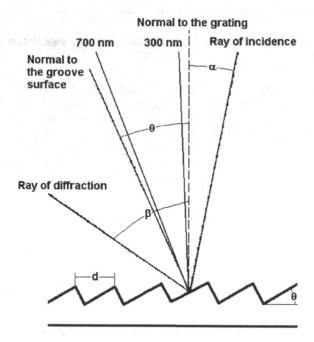

Fig. 51: Diffraction at a grating. The incident ray strikes the grating at angle α (angle of incidence). The distance d between grooves (the grating constant) will influence the angle of reinforcement and interference of the reflected light. The diffracted ray is observed at angle β (angle of reflection). Rather than showing the complete spectrum, this is represented by two wavelengths: the angle of reflection for these wavelengths at 300 and 700 nm. According to the grating equation, one finds the diffracted light of higher orders at the same angles but from different wavelengths. As a result, the wavelength of the first order for 300 nm is found at exactly the same position as the light at 150 nm in the second (order overlap). Correspondingly, one finds the wavelength 700 nm in first order, 350 nm in second order, 233.333 nm in third order and 175 nm in fourth order at the identical angle. The blaze angle θ is the angle between the normal to the grating and the normal to the groove surface

---

ⓘ   **Optical gratings for ICP-OES**

In ICP emission spectrometers, both holographic and mechanically ruled gratings are used. **Holographic gratings**, found in most types of spectrometers, are produced by applying a photosensitive layer on the glass surface and exposing this to an interference pattern. After removing the unexposed layer, the surface is etched. The stripes formed during the exposure protect the glass surface, so a fine line pattern of adjacent grooves with several thousand lines per mm results after removal of the exposed stripes.

A **mechanically ruled grating** is always made as a repeat of a "master" grating. This procedure allows exact control of the groove angle. This method of production results in gratings with a clearly lower groove density, typically some hundred grooves per mm. These gratings are usually used in Echelle optics, as for this system the groove angle is very important and the resolving power is accomplished by other means.

If a plane mirror is used in the production of a grating, one obtains a **plane grating**. If a concave mirror is used, one gets a **concave grating**. Plane gratings require additional concave mirrors to focus the spectrum onto the detector. This focusing is an inherent attribute of the concave grating.

An image of the entrance slit always occurs in the interference pattern if two wave crests overlap. Therefore, theoretically countless images exist. These multiple images of the spectrum are described as **optical orders**. At higher optical orders, a better resolution is attained. Since the spectra of different orders overlap, interfering orders must be removed with an appropriate aid (e.g. filter or pre-monochromator), as indicated in Fig. 52. For a holographic grating with a non-directed reflection, the dispersion increases with the order, while the intensity decreases. In order to obtain a light throughput which is as high as possible, these gratings are preferably used in first order.

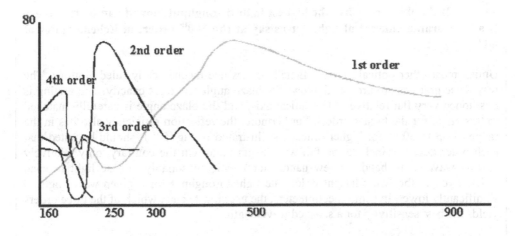

Fig. 52: Reflection efficiency of the various optical orders at an optical grating. In an optical system which uses low orders (1st to 4th orders), possible order overlaps must be removed by filters (source: Varian)

ⓘ   **What is "zero" (optical) order?**

Since a mirror is the base of an optical grating, there is a total reflection under the condition "angle of incidence = diffraction angle". This is referred to as zero order. In some spectrometers, the zero order may be used for the (automatic) basic adjustment or initialization of the optics.

The partition of the intensities among the orders strongly depends on the shape and angle of the grooves. For a strongly asymmetrical groove profile (saw-tooth profile), a specific wavelength or order is preferred. The intensity has its optimum at a certain angular position of the grating, which corresponds to a certain wavelength. This is the **blaze wavelength** of the grating and the preferred angle of diffraction is the **blaze angle**. The light throughput is optimal at the blaze wavelength and at the blaze angle. Since the majority of the ICP emission lines are in the lower UV range, the blaze wavelength typically is adjusted to 210 to 250 nm.

📖   **If the first order has the highest light throughput, how it can be that there is a measurable amount of light, let us say at the 100th order, in Echelle optics at all?**

Unlike most other optical mounts, Echelle optics use mechanically ruled gratings. The way these gratings are produced allows the blaze angle to be set exactly. The grating is positioned very flat relative to the optical axis, and the blaze angle is carefully matched to this angle for the higher orders. Furthermore, the reflection efficiency also lies in the range of up to 50 % for higher orders, as illustrated in Fig. 53. It should be noted that each order does not include the full wavelength band. On the contrary, only relatively narrow wavelength bands (a few nanometers) yield reasonably intense light in each order. Even in the two adjacent orders the light throughput for a given wavelength is significantly lower. In some spectrometers, the user must select which of the three orders yields the best sensitivity for a selected wavelength.

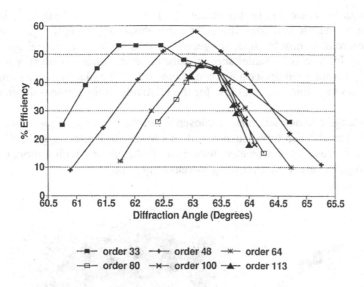

Fig. 53: Efficiency of the light diffraction at a mechanically ruled grating as a function of the groove angle (source: PerkinElmer Instruments)

Stray light arises from non-directed reflection at components of the spectrometer and from the presence of unequal distances between the grating grooves. It overlaps the diffracted light of the analyte signal and therefore can cause erroneous measurements. The stray light is essentially caused by deviations from the ideally smooth groove shape and by irregularities in the distances between grooves in mechanically ruled gratings. These irregularities are lower for holographic gratings because of the way they are manufactured. Optically coated mirrors also reduce the stray light.

3.1.3 Optical Mounts

There are different types of arranging the optical components, which are termed optical mounts. These are part of the ICP-OES instrument. Principally, two types exist: instruments that measure all wavelengths at the same time (simultaneous spectrometers or direct readers) and instruments which measure all analyte wavelengths one after another (sequential spectrometers).

Simultaneous spectrometers and some sequential spectrometers use the Paschen-Runge mount. Here the important optical elements, like the entry slit, the grating and the detector(s) are arranged in a circle (Rowland circle, Fig. 54). The grating is concave and focuses the spectra onto the detectors. In this mount, the number of optical elements is reduced to a minimum, which increases the light throughput of the optics. In the Rowland circle, a compromise must be made for geometric reasons between wavelength range and resolution. In a classic simultaneous spectrometer (direct reader) with photomultiplier tubes, the photomultiplier tubes sometimes cannot be placed at the ideal location, because neighboring lines are too close for the relatively bulky photomultiplier tubes.

The analytical wavelengths to be used must be chosen when the instrument is acquired, because replacing photomultiplier tubes in a simultaneous spectrometer is generally quite expensive. For this reason, spectrometers with array detectors and with a more or less complete wavelength coverage are becoming increasingly popular.

Fig. 54: On the left, a schematic diagram of a "double" Paschen-Runge mount is shown. All optical components lie on the Rowland circle. As an example, the array spectrometer Ciros^{CCD} is shown. On the right is a photograph of this optical system. The entry slit is at bottom left. The light strikes the primary grating at top left. There, the radiation is diffracted into the wavelengths and reaches the CCD detectors, which are arranged on the lower part of the Rowland circle. The wavelengths of an extended UV range (125 to 460 nm) are measured here. This optical system is referred to as a "double" Paschen-Runge mount, because part of the light (zero order) is reflected at the primary grating to a mirror, which acts as a virtual entry slit. A secondary grating diffracts the light of the visible range (460 to 780 nm) (source: Spectro Analytical instrument)

Sequential instruments do not have the above-mentioned restrictions with respect to line selection. Also, a better resolution is feasible here. For scanning systems, a spectrum around the analytical line can be taken to increase the security of the result. The most important mounts used for sequential instruments are the Czerny-Turner mount and the Ebert mount, variations of a very similar optical mount. While in the Ebert mount the collimator and collector is the same identical mirror, in an Czerny-Turner mount (Fig. 55) there are two separate mirrors. The typical feature of these mounts consists of rotable plane gratings at the central axis of the optics. The imprecision of the movement of the grating causes errors at the wavelength position. These must be corrected or compensated for (e.g. by periodical wavelength recalibration with a mercury vapor lamp or an emission line of the spectrum such as a carbon line) . Especially for spectrometers with these optical mounts, a signal processing technique was developed to identify the peak maximum despite possible wavelength drift (so-called "peak search" processing, see Sect. 4.2.1 "Signal Processing").

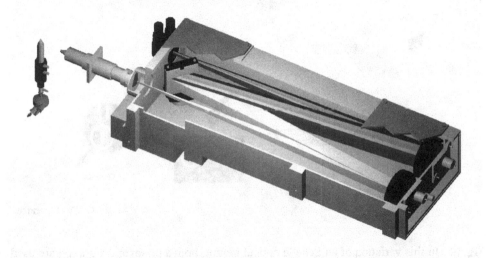

Fig. 55: In the Czerny-Turner mount shown here, the radiation emitted by the plasma (on the left) reaches the entry slit. The rays continue to the collimator, where the diverging rays are made parallel and are reflected to a grating. A stepper motor can change the angle of this plane grating to select the respective diffracted wavelength. The light is then reflected and focused by the collector. Finally, it reaches the detector (source: Jobin Yvon)

During the last few years, more and more spectrometers have been equipped with an Echelle mount [221]. In an Echelle mount [222], very good resolution is accomplished by a mechanically ruled grating with typically only about 50 to 100 grooves per mm. As pointed out in Fig. 58, it is placed very flat in the light beam [223] so that a smaller distance appears between grooves "from the point of view of the entry slit". Since the groove angle (blaze angle) is set exactly, very high orders (typically 30th to 130th order) can be transmitted with good light throughput (Fig. 53).

However, the optical order bands lie at almost the same geometric place. In the next step, the orders are sorted from each other in an axis perpendicular to the wavelength bands by a prism or a second grating (Schmidt cross-disperser, Fig. 56). As a result, a two-dimensional spectrum is generated (Fig. 57), where the wavelength bands of the different optical orders are arranged in rows one above the other.

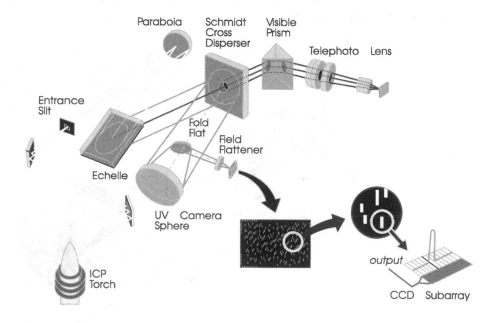

Fig. 56: In this variation of an Echelle optical mount, both a prism and a grating are used as cross-dispersing media. The prism sorts the orders of the visible range while a grating (Schmidt cross-disperser) assumes this task for the UV range. The Schmidt cross-disperser actually has a hole in it (which corresponds to the area of the fold flat mirror in front of the UV camera sphere) to allow the light of the visible range to fall onto the prism (source: PerkinElmer Instruments)

A wavelength is distributed onto three different orders. As a rule, the instrument manufacturer selects and inserts in the software or hardware which order is to be used. In some instruments the users selects which order is be used for a given analytical line. Since the optimal intensity cannot always be obtained, the sensitivity of an analytical line has to be checked in the spectrometer used.

Fig. 57: In an Echellogram, the wavelengths are positioned in bands of few nanometers, one above the other. The spectral bands are mainly in the visible wavelength range (top of the photo) (source: Thermo Elemental)

Fig. 58: In this Echelle mount, the light of the entry slit (top right) reaches the collimator (bottom right) and then passes through a prism for pre-sorting of the optical orders. A grating (top center) diffracts the light into wavelength bands, which are then reflected onto the collector (bottom left). From here, the light rays are then focused onto the detector (top left) (source: Thermo Elemental)

In the Littrow mount, collimator and collector are identical. The light ray is reflected back after diffraction at the grating on the same concave mirror. The Littrow mount typically is not used for optics of ICP emission spectrometers. However, its main characteristic, to send the ray back through the same optical elements, is used occasionally. Figure 59 shows an Echelle mount which includes the typical Littrow feature, a folded light beam.

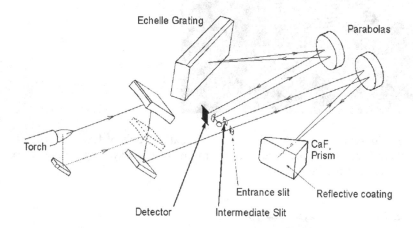

Fig. 59: Combination of Echelle and Littrow features in one optical mount. The optical order separation takes place at a reflective prism which minimizes order overlaps and stray light before the light passes onto the grating (source: PerkinElmer Instruments)

3.1.4 Light Transfer from the Plasma to the Optics

The photons emitted in the plasma should reach the detector(s) in the spectrometer with as little loss as possible. The instrument manufacturers use different strategies for this. The most frequent is to protect the optics by a quartz window. In some optical mounts, a lens is used. This protects the optical components from sample vapors originating in the plasma. In the course of time, this window can get covered with a layer of condensed vapors, so that the sensitivity decreases. Therefore, it is advisable to check the transparency of this window regularly. The simplest way to do this is by recording the intensities of several wavelengths (spread over the whole UV and visible wavelength range) over time. If a particularly strong intensity drop is observed in the deep UV, this often hints at a window that has been coated with condensed vapors. The different absorption of wavelengths in the very low-wavelength UV (e.g. at 200 nm) and the visible wavelength range is a good indicator, because the coating absorbs particularly at low wavelengths.

It should be noted that the high-wavelength UV radiation also harms the optics. The first mirror suffers from it especially. Therefore, some instruments have an automatic shutter mechanism which opens the light path to the optics only for the duration of the measurement. This helps to keep good reflection properties of the mirror over a much longer period.

Fiber optics not only protects the optics from vapors but also has the advantage that it separates the plasma and the optics from each other and allows a flexible construction of the instrument. In a spectrometer which uses several optics in one instrument, each fiber optics can be adjusted for an optimal viewing radial height of the plasma. The setting of an optimal viewing height is also possible for instruments that use direct viewing. As a rule, the light is guided to the entry slit by one or more mirrors. At least one mirror has imaging properties. In some instruments, the first mirror can be moved to select the viewing height (sometimes by software control).

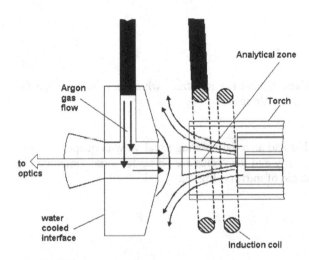

Fig. 60: One technical solution for interfacing the torch with the optics for axial (end-on) viewing. This is a water-cooled interface (a block with a hole). The hole is lined up with the analytical channel of the plasma, which strikes against this interface. Argon is blown into the hole and against the plasma flow (source: Spectro Analytical Instruments [224])

For axial viewing, technical steps must be taken to prevent either light loss due to the presence of condensed vapors in the plasma tip or the destruction of the transfer optics by the high temperatures of the plasma. A stream of argon gas that is blown in a direction countercurrent to the plasma flow can prevent this. In many spectrometers, this is achieved using a design similar to that found in ICP-MS, as in the example shown in Fig. 60. This interface is a cone with a hole in the tip, the hole being aligned coaxially with the analyte channel. The plasma strikes this cone, as illustrated in Fig. 61, and an argon flow then removes sample vapors carried by the analyte channel. Of course, this cone must be cooled intensely. The advantage of this design is the fact that the greatest possible length of optical path is purged with inert gas (to remove absorbing oxygen), so that transparency is optimal in the vacuum-UV range. When shear gas is used to cut off the plasma tip, reduced transparency in the vacuum-UV range must be accepted. The shear gas jet is positioned perpendicular to the analyte channel and off its axis. The plasma tip is blown away with a high flow of air from a wide jet. This air flow also serves to remove the waste heat. The tip of the plasma is removed in all cases because this is the recombination zone and would otherwise act as an absorptive layer, which would accentuate matrix influences and limit linearity.

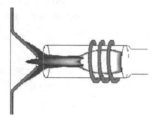

Fig. 61: The horizontal plasma strikes against the interface, which is positioned on the same axis as the analyte channel (source: Varian)

Some instruments allow both radial and an axial viewing (dual view concept). In these instruments, the plasma lies horizontally and the optics are set up for axial viewing. The radial viewing is then done with the aid of mirrors (Fig. 62) [225].

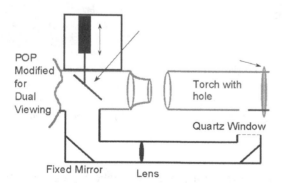

Fig. 62: Example of transfer optics that allow both radial and axial viewing. The plasma is contained in an extended torch with hole in the lower part. The light from the plasma for radial viewing is reflected with two mirrors (one is labeled as such at bottom left and the other is at bottom right) toward the spectrometer (top left). Alternatively, the plasma can be viewed axially by means of a mirror (indicated by the arrow, top left) (source: Thermo Elemental)

In the wavelength range below 190 nm, oxygen from the ambient air absorbs light. The absorption coefficient for oxygen rises particularly steeply below 190 nm, as illustrated in Fig. 63. In order to reduce absorption in the wavelength range below 190 nm, the oxygen must be removed from the light path. This is done by either purging the optical path with inert (oxygen-free) gas or evacuating the optics (hence the name vacuum-UV). The solution by evacuation is more expensive technically, and this will affect the price of the instrument. Therefore, quite a number of instruments use a purge gas, although in this case a gas tight instrument must be specified in order to exclude oxygen from entering the optics as much as possible.

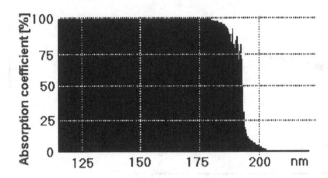

Fig. 63: Absorption coefficient of oxygen

✗   Can I also use argon as a purge gas for the optics?

In order to displace the oxygen of the ambient air from the optical path, the entire path length from the plasma to the detector is purged as completely as possible or filled with an oxygen-free gas (purge gas). The choice between nitrogen or argon is influenced not only by analytical considerations but also by cost. The frequency of usage, the gas installation, the price of the rental of the gas cylinders and the gas costs can influence the decision in favor of either gas.

✂   **I observe a drift in the vacuum-UV. What can the reason be for this?**

As a rule, a specific time must be allowed for all the oxygen to be displaced from the spectrometer. It is recommended that this time should be determined experimentally for the particular instrument used. To exclude other influences, it is suggested that this should be done using a pair of wavelengths of the same element with one clearly below and the other clearly above 190 nm. The ratio of the intensity below 190 nm to the intensity above 190 nm is calculated, and this ratio is then plotted as a function of time. Figure 64 shows this using the example of phosphorus measured at 178 and 213 nm.

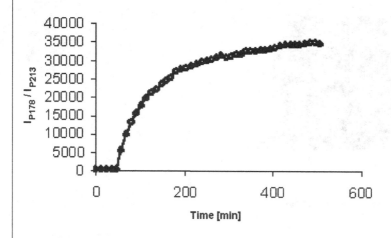

Fig. 64: Change in the transparency of the optics in the vacuum-UV on purging with nitrogen. The figure shows the ratio of the intensities of the phosphorus lines at 178 nm and 213 nm measured during the initial purging of the optics of a PerkinElmer Optima 2000

✂   **Although I have purged as completely as possible with an inert gas, sensitivity is poor in the vacuum-UV**

Even tiny quantities of oxygen can cause the transparency to decrease quite noticeably in the wavelength range below 190 nm. Minute leaks in the purge gas tubing could allow enough oxygen to diffuse into the purge gas to make the transparency of the optics substantially lower than it would be with extremely pure purge gas. Considerable differences can emerge when comparing different tube materials that supply the purge gas to the spectrometer. Sometimes tube material, although it looks smooth and solid to the bare eye, is very porous at microscopic level. Enough oxygen can penetrate through these invisible holes to markedly reduce the sensitivity of the analytical line. Therefore, invisible pores in the tubing material (and defective quartz windows in the transfer optics) should be considered as possible reasons for poor light throughput. Copper tubes are to be preferred over synthetic materials as these are far less permeable to gases.

Caution! Do not use all the automatic functions without considering the consequences!

In some instruments, control of the purge gas flow is tied into the analytical method. If the user needs to purge with inert gas by a specific method, then it seems to make sense to trigger the purge in the method stored in the software. However, since a stabilization time is required, but the software may not make any provisions for it, a drift can be produced by this way of operating the ICP-OES system (compare also box: "I observe a drift in the vacuum-UV. What can the reason be for this?") It happens quite frequently that one method turns the purge gas flow off (because it is not needed) while the next method turns it on again without the required stabilization time.

Note: Measured values of emission wavelengths well above 190 nm are unaffected if these are measured with optics purged with an inert gas.

3.2 Detectors

Electronic components and detectors have simplified and accelerated the transition of qualitative spectroscopy to quantitative spectrometry [226, 227, 228]. In the early days of OES, then mainly using spark excitation, the spectrum of a sample was recorded with a photographic plate (Fig. 65). The user rapidly gained a qualitative impression of the components in the sample using that procedure, since several lines of an element could be found. As the presence of several lines could be checked immediately, a false interpretation was unlikely after some experience in working with the technique. A quantitative determination was, however, quite difficult, since the extent of blackening of the exposed areas could be only imperfectly determined.

Substitution of the photographic plate by photomultiplier tubes led to the wide use of emission spectrometry by a large circle of users because the emission signals could be processed much more easily. However, a disadvantage also became evident very quickly. The great advantage of the photographic plate had been the extensive spectral information. The analyst received information not only about the analytical line but also about other spectral lines due to the analyte and about the spectral surroundings of these lines. The information content of the photographic plate was sacrificed in favor of a simple processing system on changing to photomultiplier tubes [229].

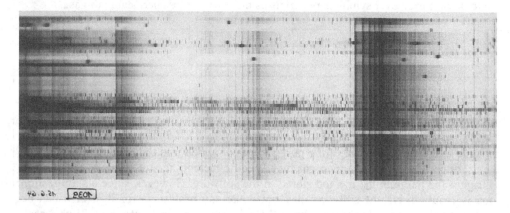

Fig. 65: Prior to electronic detectors, spectra were recorded with a photographic plate. The spectra of different standards and samples are seen here one above the other. The intensity is taken from the shades of gray and black

3.2.1   Photomultiplier Tube (PMT)

The photomultiplier tube (PMT) converts photons into electrons by the photoelectric effect at a photo-cathode [230]. The electrons released are accelerated in the vacuum by another electrode (dynode), which has a higher potential than the first one. The electrons strike its surface and further electrons are released. In subsequent dynodes with steadily increasing potentials, an electron avalanche is produced. The resulting electric current is used as a signal. Figure 66 illustrates this.

Even if no light falls on the detector, a few thermal electrons are released. These electrons form the dark current, the magnitude of which depends on the temperature. The dark current limits the detection limit in those cases where the background emission from the plasma is so small that the noise is recorded by the detector.

The higher the acceleration voltage between the dynodes, the more electrons are produced. Thus the sensitivity can be regulated within certain limits. Typical voltages are in the range 300–1 000 V. In parallel with the enhancement of the sensitivity of the detector with increasing voltage, the dark current is also elevated. Therefore no benefit will be observed with respect to limits of detection. The linear working range of a photomultiplier tube typically covers six orders of magnitude.

Photo-cathode material (mostly alkali-metal alloys) and transparency of the window determine the sensitivity, which also depends on the wavelength. There are special detectors which use only light within a certain wavelength range (e.g. "solar blind" detectors, which detect light only in the UV wavelength range) .

Depending on the geometric arrangement of the window and the photo-cathode in the photo tube, photomultipliers are divided into two categories: a "side-on" tube where the light comes from the side and a "head-on" tube with the light comes from the top of the tube.

Fig. 66: Schematic diagram of a photomultiplier tube. Light falls from the right through a window at top right. At the first photo-cathode (marked 1 e-) the first electron is released, and this is multiplied by a cascade of electrodes which have increasingly higher voltages. A "side-on" multiplier is shown in this figure (source: Jobin Yvon)

3.2.2 Solid-State Detectors

The restriction of a photomultiplier tube to the information at only one point meant that virtually "blind" measurements became usual without the chance to assess the signal. Occasionally, this led to seriously erroneous results. In order to recover at least part of the spectral information content during routine analysis, a scanning ICP emission spectrometer would be required. However, scanning over the spectral vicinity around the analyte line takes a lot of time and diminishes the precision of the analysis [231]. Therefore, this type was used by only a few ICP-OES instruments available on the market, although it makes sense to enhance the security of the result.

The two-dimensional picture which is produced by Echelle optics is optimally suitable for the construction of a solid-state detector that yields spatial information. The material of these detectors is a photosensitive semiconductor as used in video cameras. Since the requirements of the video camera and ICP-OES clearly differ from each other, these electronic components were developed further to adapt them to the requirements of ICP emission spectrometry.

Solid-state detectors register wavelength bands and allow a sizeable gain of spectral information (array type) combined with simple signal processing capabilities. Utilization of modern electronic devices leads to more spectral information and thus security of the results. Furthermore – thanks to the simultaneous measurements – higher speed and precision of the analysis is gained [232, 233]. The use of array detectors has several analytical advantages for ICP emission spectrometry [234]. Only the use of solid-state detectors enables the simple quantitative signal processing of the photomultiplier tube to be combined with the vital spectral information density of the photographic plate [228].

The smallest pictorial unit of a solid-state detector is the **pixel**. The pixel width corresponds to the illuminated part of the exit slit of a photomultiplier tube system. The pixel height is set to the slit height in some instruments, while in other spectrometers the image of the slit height is composed of different pixels lying one above the other. In a photodiode-array, a diode corresponds to a pixel. The band of neighboring pixels which covers the complete detector width is described as an **array**, while a **subarray** covers only a part of the detector width. This subarray can be defined for one measurement only (Fig. 67) or can be permanently installed as an area on the detector during production.

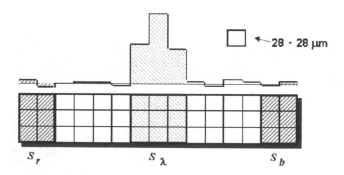

Fig. 67: A subarray is a group of adjacent pixels (in this example the pixel is a square with an edge length of 28 μm) on a solid-state detector, here a charge-injection device (CID). The subarray consists of 3 rows and 15 columns. S_λ is the part of the subarray which is used for processing the peak intensity; S_r and S_b are the two areas which are used for background correction (source: Thermo Elemental)

The **charge capacity** of a pixel is a measure of how many electrons can be collected in a pixel without being lost to surrounding pixels or electrodes. The charge capacity determines the dynamic range and linear range of the detector. If the charge capacity is exceeded, the charges spill over to adjacent pixels. This is commonly described as **blooming**; which appears when the solid-state detector is strongly overexposed. Blooming is reversible by reading out the charges from the detector. To avoid blooming, the pixels are read out at different measurement frequencies, or other technical measures, such as simple grounding, can be taken to prevent the spill-over of particularly high amounts of charges.

The noise of a detector is essentially caused by the charges generated by photons. These are read out by the detector and lead to the **read-out error**. In a CID, the read-out error can be statistically averaged by a multiple read-out of the same pixel and therefore minimized.

All solid-state detectors are categorized under the term **charge-transfer device** (CTD) [235]. The CTDs consist of doped pure silicon. This light-sensitive material produces charges when struck by photons. (Often the term "electrons" is used despite the fact that a positive charge is generated.) CTDs have relatively high quantum efficiencies and thus a high light sensitivity (Fig. 68). In addition, they have a relatively low dark current. This low dark current is further reduced by external cooling to typically below 0 °C. At these temperatures, the detector may ice up by water vapor condensing from the air. The instruments on the market are equipped with different technical solutions to avoid the icing up. In some instruments, the detector is permanently purged with argon gas. The low dark current has a great advantage, particularly in the lower wavelength range, since there is very little background emission of the plasma. In this case, the precision of the measured signal is not deteriorated by the dark current of the detector.

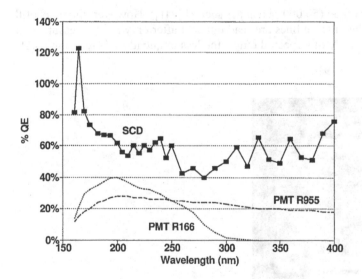

Fig. 68: Comparison of the quantum efficiencies (QE) of different detectors used in ICP emission spectrometers. Two types of photomultiplier tubes whose sensitivities cover different regions of the spectrum are compared with a segmented charge-coupled device detector (SCD). The SCD represents the group of the solid-state detectors. A quantum efficiency of over 100 % can be observed if one photon releases more than one charge on a statistical average (source: PerkinElmer Instruments)

A **photodiode array** (PDA) is a solid-state detector where the created charges are conducted by an addressable transfer line of the single photodiode to the preamplifier. The adjacent diodes are arranged in a row (array). Photodiode arrays have a relatively high dark current (approx. 500 c/s per pixel) and a high read-out noise. These characteristics make their utilization in commercial ICP emission spectrometers unfavorable. However, before commercial array-spectrometers appeared on the market, research instruments with PDAs were described in the literature [236, 237, 238].

In the **charge-injection device** (CID, Fig. 69) , the charges can be addressed by a combination of columns and rows and are measured by a preamplifier. Read-out is then repeated. Thus, the read-out error is reduced by averaging several measurements. This read-out strategy is adjusted to suit to the detector. In the CID, the increase in the charges is registered constantly as is illustrated in Fig. 70. (Thus transient signal processing is an inherent feature of this type of detector, cf. Fig. 71). The duration of the measurement is extended as much as possible without exhausting the charge capacity. By this strategy, it is possible to obtain the best signal/noise ratio [239]. The measurement process must be terminated before the charge capacity is reached since otherwise charges would be lost and the measurement could not be utilized. Therefore, a threshold value is defined (approx. 80 %). If the charges in a pixel reach this value, these charges are read out and the elapsed time is noted [240]. From the ratio of charge ["counts" = "electrons"] to time [s] the intensity is calculated [counts per second; c/s or cps].

The measurement takes time (50 000 pixels per second [241]). However, the pixels fill at different speeds, and the intense lines are read out first after only a few measurement cycles. The lower the intensity of a spectral range, the less frequently it is read-out [242].

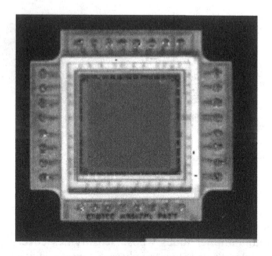

Fig. 69: Photograph of a charge-injection device detector (source: Thermo Elemental)

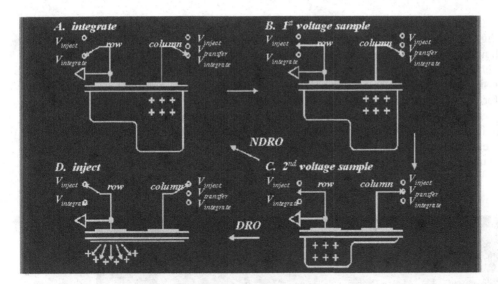

Fig. 70: Schematic representation of the read-out process of a CID detector. The figure should be read in a clockwise direction: **A** At top left, the production of charges (+) by photons striking the detector is shown. The charges are retained in a potential well. (The potential is illustrated by the line below the detector surface.) **B** The number of charges is then read out. **C** Next, the potential well is moved (to the left), which shifts the charges. The number of charges is then also read out there. The circle A → B → C is termed non-destructive read-out (NDRO). **D** Only if sufficient charges have accumulated for good counting statistics is the potential lifted altogether, and the charges are then dispersed to make room for a totally new measuring cycle (source: Thermo Elemental)

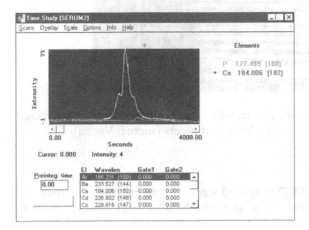

Fig. 71: Because of the read-out characteristics of the CID, each signal is recorded as a transient signal. This can be utilized for some special applications, e.g. ETV coupling (source: Thermo Elemental)

The **charge-coupled device** (CCD) is constructed of rows of pixels, as shown in Fig. 72. In the CCD, the charges are moved in rows to the read-out electronics. The longer a row is, the bigger is the read-out error. There are several strategies for the read-out: Most frequently, the array with the shortest read-out time of the system is exposed. Using this pre-measurement, the software then determines an optimal measurement time for the subsequent analytical measurements for every array or subarray to obtain the best possible signal/noise ratio. Again, the ratio of the charge to the measurement time gives a figure for the intensity. Alternatively, multiple exposures are carried out and the measurement with the best signal/noise ratio but well below the saturation of the detector is taken as the signal reading.

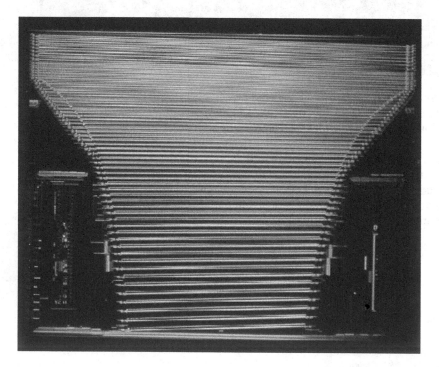

Fig. 72: Photograph of a charge-coupled device, which has been optimized for the image of an Echellogram. The rows correspond to the optical orders (source: Varian)

The charges are read out from a CCD in several steps. For the read-out, the rows are further subdivided according to height as well as along the wavelength axis. In addition to the light-sensitive row of pixels, two more rows, which are not light sensitive, exist for the read-out process (Figs. 73 and 74). Photons are converted into charges in the upper area (A register). These move to the B register because of an electric potential slope, and remain here (an intermediate memory) until the read-out of the charges is triggered. Then A and B registers are separated by a potential barrier, while the charges flow rapidly from the B register into the C register because of a steep potential slope. Immediately thereafter, the latter two registers are separated again. Now there is plenty of time for the read-out electronics to read out the C register pixel by pixel, from where the signals are ultimately transferred to the software. Figure 74 clarifies this multistage event.

In order to avoid blooming, the **segmented charge-coupled device detector** (SCD) is subdivided into relatively small photosensitive subarrays [243]. These subarrays are formed during manufacture. Each subarray has a grounding wire around it to efficiently prevent any potential overspill of charges to neighboring subarrays. Each subarray can be addressed individually by the software. For read-out, groups of subarrays with identical measurement times are formed, and these are measured simultaneously. The user system can force the system to always measure simultaneously, which reduces the operating range to 4.5 orders of magnitude.

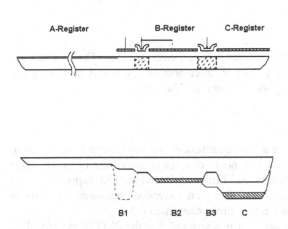

Fig. 73: Cross-section of an SCD (segmented charge-coupled device detector). The upper part of the figure shows the electrodes, which are partly embedded (alongside in the A register) and which partially cover the surface (B and C register). The areas over the B and C registers are covered since they are needed only for the read-out process. In the lower part of the figure, the potential diagram is shown. Charges are released by photons in the A register, which move, driven by a low potential slope, to the B register, which serves as an intermediate memory. A potential barrier (B3) is set up between registers B and C, which ensures that the charges remain in the intermediate memory until they are read out. For the read-out process, the potentials are switched for a short time (marked by hatched area), so that the potential of the C register now is lower and the charges in there flow very fast from B2 to C. A temporary register (B1) is installed in order not to lose the charges which are generated during this time in the A register. As a last step, the charges are then read out one after another from the C register (source: PerkinElmer Instruments)

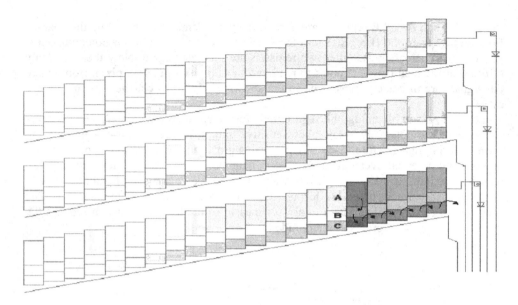

Fig. 74: The registers of A, B, and C are arranged one over another. The charges are generated in register A from where they flow to intermediate register B until they are finally read out sequentially through register C (source: Varian)

Charge-transfer devices typically have a high quantum efficiency (about 3 times that of a photomultiplier tube) and an extremely low noise characteristics of the detector. This has an immediate influence on the instrumental limits of detection. The improvement in the limit of detection theoretically expected from an extended measurement time is actually observed throughout the measurement times examined [244].

Generally, the working range of any detector is restricted. For the PMT, it is controlled by the high voltage applied (typically 300–1000 V). The working range of a solid-state detector is limited by the charge capacity. In order to extend the working range, the measurement and/or the read-out times are varied. In the CID, the total measurement time is fixed. Within this period, the detector is read out repeatedly up to a threshold value (to prevent the charge capacity from being exceeded). In the CCD, the read-out time is varied in such a way that the charge capacity is not exceeded and a sufficient count rate results (Fig. 75). The measurement time to collect charges may be very short at very high intensities (e.g. in the millisecond range). Since these short measurement times would cause poor reproducibility, short read-out times are repeated until a longer total time is reached, giving better reproducibility. The user can set these times. The software then determines the most favorable combination of read-out and total measuring time within this frame.

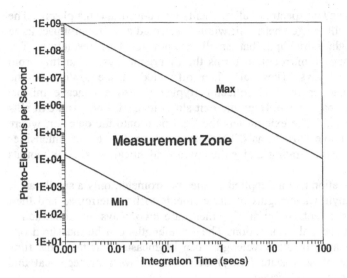

Fig. 75: The working range of a solid-state detector (here the example SCD is taken) is 4.5 orders of magnitude. It can be increased by changing the read-out times. At a given intensity (axis marked "Photo-Electrons per Second") the measurement time (here marked "Integration Time") is varied within the area marked "Measurement Zone". Within this measurement zone, the system will utilize the optimal working range of the detector. Below the minimum zone (Min), the intensity is too low. Above the maximum zone (Max), saturation is observed (source: PerkinElmer Instruments)

3.3 Types of Emission Spectrometer Mounts

The spectrometer will, by means of the optics, separate the analytical line from the emission originating from the plasma and measure the corresponding intensity with the detector. In principle an emission spectrometer consists of the following components:

- Radiation source (plasma)
- Entry slit
- Collimator
- Grating
- Collector
- Exit slit
- Detector
- Signal processing.

The radiation source emits line spectra of all elements introduced into the plasma. The light radiates to the entry slit. Its geometric slit width is selected as a compromise: large enough to give good light throughput but small enough for best resolution. The collimator (typically a concave mirror) transforms the divergent rays originating from the entry slit into parallel rays. These are then diffracted at the grating or the combination of grating and prism. The collector (typically also a concave mirror) focuses the light onto the image plane. Here, the exit slit is located. Entry and exit slits typically have the same width. The exit slit lets the light pass onto the detector, where the radiation is converted into electric signals. These are then transformed into intensities by the signal processing unit. Grating and collector are identical in special mounts (Paschen-Runge) .

Of all the spectral information that is supplied by the spectrometer, only a small part is of prime interest: the analyte wavelengths of the elements to be determined and their immediate spectral environment. In principle, there are two ways of constructing spectrometers to gain the spectral information: The wavelengths can be measured one after another (monochromator) or all wavelengths can be measured at the same time (polychromator). The use of solid-state (array) detectors allows new technical and analytical solutions for each type of system.

3.3.1 Monochromators

Monochromators are used for sequential measurements. Their main advantage is the flexibility of wavelength selection. Typically, the diffracting element is a plane grating, which is rotated to direct the different analyte wavelengths onto an exit slit/detector combination. The grating is moved by a software-controlled stepper motor. The optical mount most often used is the Czerny-Turner mount. A few sequential instruments use Echelle mounts. As a rule, monochromators are designed in such a way as to give better resolution than polychromators. Scanning monochromators can generate spectra in the vicinity of the analytical line, which enhances the security for an accurate determination.

3.3.2 Polychromators

For simultaneous determinations, polychromators (direct readers) are used. Entry slit, grating, and up to 60 exit slits and the corresponding detectors (channels) are fixed permanently on the Rowland circle. As a rule, a concave grating is used as the diffracting element, and this diffracts and focuses at the same time. The Paschen-Runge mount is almost exclusively preferred for simultaneous instruments. Even if all the analytical lines are measured simultaneously, the background correction measurement is typically carried out in a sequential manner: most instruments have a rotatable quartz refractor plate, so that several points are measured one after the other, and a sequential background correction is given. The main advantages of classical simultaneous spectrometers consist in the stability (short- and long-term) and the high analysis speed. At the time of the purchase of the instrument, the user must select which elements will be measured at which wavelength. A later change is laborious and therefore quite costly [245].

3.3.3 Array Spectrometers

The option to measure spectral ranges simultaneously with photosensitive solid-state detectors generates a new class of ICP-OES instruments, the array spectrometers. Although not mandatory, array spectrometers are frequently equipped with Echelle optics. Here very good resolution is possible with good light throughput. Because of the relatively short focal length, it is possible to design compact instruments which are wavelength stable. Therefore, many manufacturers design their instruments on the basis of an Echelle mount [246].

Echelle optical mounts can deliver very good resolutions. Since Echelle optics use different orders, the resolution also changes, depending on the optical order and consequently the wavelength. Since the lowest wavelengths are measured at the highest orders, the resolution is the best in the low UV. This characteristic fits well with the spectra generated in ICP emission, where the highest density of analytical wavelengths appears in the lower UV range.

4 Method Development

In method development the goal is to choose parameters which produce a clear and undisturbed signal for the analytical measurement. In ICP-OES, the analyte signals are located on a sometimes high background, which originates from the plasma radiation. A signal that can be processed unambiguously will be obtained only if the signal stands out clearly from the background, which requires a large signal/background ratio.

All measured intensities are accompanied by random fluctuations, also described as noise [247]. There are various reasons for these noise components: pulsations of the peristaltic pump, irregularities during nebulization and in the nebulizer chamber, flicker of the plasma (partly caused by fluctuations of the RF generator power or the gas flows), poor wavelength stability, fluctuations at the detector and the subsequent signal processing, and sometimes variations in the electric power supply. The fewer and less pronounced the noise components are, the better is the reproducibility of the result. The reproducibility may be used as a performance criterion for an adequately intense signal. For a very small signal which can hardly be distinguished from the background noise, the concept of the limit of detection is used. Finally, the working range should be defined.

In ICP-OES, similarly to other analytical techniques, systematic or random deviations of the measurement from the true value are observed which are a result of interference. As a rule, this interference is caused by components of the sample (matrix). In order to avoid erroneous measurements, the user selects operating procedures for the analytical method. This must be developed before beginning the series of routine measurements for a given application. The more care used for method development, the safer will be the results of the subsequent routine measurements. Of course, a method can be used again later for comparable applications.

In the course of method development in ICP-OES, the measurement parameters which are used in the analytical measurement program of the spectrometer are specified. This is a part of a more extensive method specification which includes all operating steps starting with sampling, sample preservation and sample pretreatment, all the way to the documentation and interpretation. For analysis with ICP-OES, a number of parameters are selected and optimized, the most important of these being

- Analytical lines
- Excitation conditions
- Measurement times and repeats
- Selection of the processing techniques (particularly the background correction)
- Checking for and correcting for non-spectral interference.

In the method, the user selects which wavelengths (analytical lines) shall be used for the quantification. The choice of the analytical lines is substantially influenced by the characteristics of the spectrometer, particularly resolution and utilizable wavelength range. As stated in Sect. 3.1.1 "Resolution", good separation of neighboring lines is important, and a spectrometer with a better resolution will have more potential lines available. Furthermore, the utilizable wavelength range has to be considered. Depending on the spectrometer used, constraints can arise here.

As described in the Chapt. 2, the excitation conditions in the plasma have a substantial influence on the intensity of an emission line. In individual cases, the intensity ratio between the analyte line and the plasma emission can be increased by modifying the plasma conditions so as to minimize spectral interference, enabling an emission line which was previously heavily interfered with at other excitation temperatures to be utilized analytically [248]. In Fig. 76, the result of the optimization of the excitation conditions to minimize spectral interference is shown. Thus, the optimization of the excitation conditions may be an important step during the method development. This means that the user should optimize the RF generator power and the argon gas flows, particularly the nebulizer gas flow. For the radially viewed plasma, the viewing height has to be optimized also.

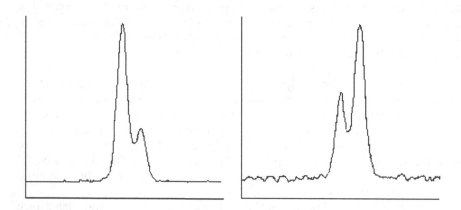

Fig. 76: By changing the excitation conditions, the intensity ratio between analyte and interferent can be changed quite markedly in some cases. In this figure, the interference of the Cd line at 228.802 nm (at a concentration of 1 mg/L – the left hand peak in each spectrum) by As (at 10 mg/L – the right hand peak) is displayed. The left hand spectrum shows the result of the optimization for Cd. The excitation conditions are: RF power of 1 280 W and a viewing height of 22 mm. The right hand spectrum shows the optimization for As with the instrument settings: RF power 600 W, viewing height 6 mm (generated with a PerkinElmer Plasma II)

Furthermore, the method specifies how the measurements have to be carried out. The user selects the measurement times and the number of repeat measurements. Both parameters are mainly reflected in statistical criteria such as reproducibility and are necessary for establishing analytical performance criteria. Since ICP-OES is a fast technique, multiple repeat readings can be obtained without a great time penalty.

Consequently, several repeat readings should be taken. The overall analysis speed is frequently influenced by the delay time to give a steady-state signal and the rinse time to avoid carryover, which is typically longer than the time for a repeat measurement. Therefore, it is strongly recommended to measure repeatedly in order to get meaningful statistical information [249].

Carryover can be reduced by increasing the delay time (before the beginning of the measurement of the sample or calibration solution) or the intermediate rinse time (between the samples with a separate rinse solution) to an acceptable value.

Depending on the spectrometer used, the user can define in the method how the signal registered by the spectrometer shall be converted into the net intensity value. In this instance, the user defines how the peak or its maximum shall be registered and calculated and in what way the background shall be corrected.

Besides spectral interference, non-spectral interference can also occur. Its influence must be determined and corrected or compensated for. For example, non-spectral interference is observed, if the solutions have different viscosities. Then the composition of the calibration solution or the sample pretreatment must be matched or an internal standard must be used.

4.1 Wavelength Selection

4.1.1 Working Range

The first criterion for the choice of the analytical line is the range in which the concentration of the analyte (the element) is anticipated. This means that the user should have an idea of what the concentration will be. Because of the wide linear range (typically 5 orders of magnitude), a vague estimate is usually sufficient. However, even this range has its limits.

If trace analysis is to be carried out, the most sensitive lines have to be used. The wavelength tables normally used include various parameters concerning the intensity. The following are common:

- Background equivalent concentration
  (see Sect. 4.1.1.1 "Background Equivalent Concentration (BEC)"):
  Here, the *smaller* the numerical value, the more sensitive the line.

- Signal/background ratio and

- Intensity

  For these two parameters, the *bigger* the numerical value, the more sensitive the line.

It should be noted that for a signal/background ratio or an intensity entry it is necessary to include the concentration of the element at which the value was determined. This is an essential part of the information. Unfortunately, in several tables these entries are missing. Furthermore, it has to be borne in mind that, because of differences between instrument parameters, intensities may be compared with each other only within one set of tables!

If higher concentrations are expected, the working range should be estimated to enable a less sensitive line to be found such that the anticipated concentration falls into the probable working range of that line.

4.1.1.1 Background Equivalent Concentration (BEC)

Since the signal intensity depends on the concentration, a bare intensity value is not meaningful in the wavelength tables. The concentration of the element at which the intensity was measured should always be included. This manner of the representation is somewhat complex. Therefore, a performance criterion that includes both intensity and its corresponding concentration has been defined in OES: the background equivalent concentration (BEC):

The background equivalent concentration is the elemental concentration required to produce an analyte signal with a same intensity as the background signal.

ⓘ   **The smaller the BEC numerical value, the more sensitive the line**

If a small concentration of the element is sufficient to produce a signal of the same level as the background, then this is obviously better than if larger quantities were necessary.

The definition includes a kind of "reciprocal sensitivity" . In addition to this, the background is included as a determining factor.

The calculation is carried out using the following equation:

$$BEC = I_{background} / I_{net} \cdot c_{analyte}$$

where $I_{background}$ is the intensity of the background measured with the blank solution or next to the analytical line, I_{net} is the difference between the total intensity of the line and the intensity of the background and $c_{analyte}$ is the concentration of the analyte that yielded the net intensity.

At first sight, the background equivalent concentration appears to be an abstract quantity that does not correspond to any analytical performance characteristics used in other techniques. On closer inspection, it contains the reciprocal signal/background ratio and the reciprocal sensitivity (minus the background contribution). In addition, it is tied to the limit of detection by the equation [250, 251]

$$c_{LOD} = RSD_B \cdot BEC \cdot k$$

where c_{LOD} is the limit of detection, RSD_B is the relative standard deviation of the background (RSD of the blank reading without background correction applied) and k is the statistical factor which is used for calculating the limit of detection. This equation can also be used for calculation of the limit of detection. The advantage of this procedure is the fact that the two components which decide the limit of detection, the sensitivity and the noise, are measured independently. If the limit of detection is worse than anticipated, it will become obvious what the cause is: high noise (RSD_B) or poor sensitivity/high background (BEC) [252].

In addition, the equation can be used for a first estimate of the expected order of magnitude of the limit of detection. If one inserts a typical value of 1 % (0.01 as decimal number) for the noise and the fairly common value of 3 for k, the equation will change to:

$$c_{LOD} = 0.01 \cdot BEC \cdot 3 = 0.03 \cdot BEC$$

This is a very rough estimate. The assumption of 1 % is conservative if one considers the fact that the relative standard deviation of the background without correction is determined. Since the background emission can be very intense, particularly at higher wavelengths, this condition will be satisfied with most spectrometers. However, the user should make sure that this assumption is also true for his or her instrument and, if necessary, insert another numerical value.

This rule of thumb for the first rough estimate of the expected order of magnitude for the limit of detection should really be used only to get a rough idea for line selection with respect to sensitivity. At the present time, the ICP-OES wavelength tables known to the author include the BEC results for radial viewing. To estimate the limit of detection for axial viewing, the value should be divided by the factor 10.

Both the intensity of the signal and the intensity of the background depend on the plasma temperature. While the intensity of the analyte signal is highest at the norm temperature and becomes lower at any deviation from it, the intensity of the background increases with an increasing plasma temperature. Therefore the signal/background ratio changes, and with that the background equivalent concentration.

4.1.2 Freedom from Spectral Interference

Spectral interference is one of the main causes of erroneous results in ICP-OES. It originates from emissions of the structures of the spectral background and from sample components which emit light in the immediate neighborhood of the analytical line.

When choosing an analytical line, it is useful to obtain as much information as possible about the sample beforehand. The more that is known about the sample matrix, the faster the line selection will be. As well as an estimate of the concentration range to be expected for the analytes, as much knowledge as possible of the matrix components is important. It is particularly important to be able to estimate which accompanying elements could be present and have an idea of their concentrations. If the information cannot be obtained elsewhere, it would be advisable to perform a screening analysis before starting the line selection.

The strategy of a screening analysis is to measure 20 or more elements very rapidly. In contrast to the usual procedure of method development, where the goal is to select one line, several lines per element are utilized. During or after the analysis the spectra are reviewed, and the results are critically compared to select a plausible value for each element [253, 254]. This procedure can even be refined by the application of multivariate processing techniques [255].

One should also consider the instrument specifications, particularly the resolution, as factors determining the choice. Often the resolution of the spectrometer decides whether there is a spectral overlap or not.

The distance between two of neighboring lines has a substantial impact on how severe a spectral interference will show. This distance can be expressed as a multiple of the resolution. To restate: the analytical line is regarded as interference-free if the signal of the interferent is approximately equally large and is at a distance of one resolution. In this criterion, a further important statement is already included: the intensity ratios are decisive for judging the suitability of a line in the presence of an interfering line. If the distance does not change, then the impact of a possible interference becomes less severe if the intensity ratios between analyte and interferent shift in favor of the analyte. Thus in special cases a wavelength may utilizable although the distance between analyte and interferent lines is less than one resolution. On the other hand, a line can prove to be useless even at a sufficient distance to the interfering line if there is an unfavorable analyte/interferent ratio.

📖   **Sometimes I see particularly wide lines**

Lines which show a natural FWHH of significantly greater than 10 pm are displayed in their natural form by spectrometers whose resolution is much lower. These relatively wide lines are the result of a hyperfine structure which is not completely resolved. This is observed mainly for rare earth elements, whose f-orbitals take part in the structure of the electron shell.

Therefore, this situation is restricted to some special applications.

As a basic rule, it makes sense to check several wavelengths per element for their potential suitability [256].Depending on the instrument used, the strategy will vary. With classical simultaneous spectrometers, the available channels are fixed in the instrument. Here the method development consists essentially of a check to see whether the available wavelengths are appropriate for the working range and whether there is spectral or non-spectral interference (and to look for ways of correcting this if necessary). If a classical sequential instrument is used, the flexibility of the system of line selection should be utilized, and an undisturbed line should be looked for as described below. The same routine for line selection also applies to array spectrometers.

Since it takes a great deal of time with a sequential instrument to generate all the necessary spectra for judging the suitability of a wavelength, one might be tempted to "streamline" a thorough method development by using only a small number of wavelengths and samples. This procedure makes it important to choose samples which are considered representative for the whole series [257, 258]. Also the evaluation of the wavelengths takes time, so it may be a good idea to start with a test to exclude obviously unsuitable lines first. Of course, all boundary conditions should be taken into consideration here. When trying to use the information gathered during this test, the conditions elsewhere that led to selection of a wavelength should be considered and compared to those relevant to one's own application. As an example: wavelengths which may be usable with a spectrometer of good resolution may have been described as unsuitable with spectrometers of poorer resolution [259].

ⓘ   **Why do "new" analytical lines sometimes appear in methods developed using modern spectrometers?**

The process of selecting an analyte wavelength includes considering the experience of former experimenters. It makes a lot of sense not to re-invent the wheel every time someone starts a method development. However, much experience from the early days of the ICP becomes outdated from the point of view of modern spectrometers. Hence, the set of wavelengths used in the early days sometimes requires revision, because with spectrometers of better resolution, more wavelengths qualify as potential analytical lines than was the case with the spectrometers of the late seventies. Hence, some wavelengths which were classified as unusable a long time ago are now used analytically. Examples are the P line at 213 nm (with its interference by Cu) and the Ni line at 221 nm (interference by Si, cf. Fig. 77). In a spectrometer with a good resolution, these lines will not interfere in many applications, and these wavelengths can be meaningful alternatives. In addition to a literature search for successful applications and a consultation with colleagues, the critical user should therefore perform an experimental check of all the wavelengths available, also including those that hitherto had been known only from some kind of historical records.

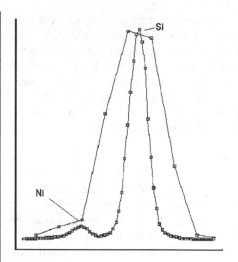

Fig. 77: The use of a spectrometer with superior resolution makes possible the selection of new lines including those wavelengths which were considered to be unsuitable because of reported spectral interference. The figure shows the spectrum of the Ni line at 221 nm in presence of an interfering Si line. Both spectra were recorded from the same solution on the identical instrument (Optima 3000), which was set to different resolution settings: the actual resolution is 7 pm for the narrow peak and 23 pm for the wide peak. With the better resolution, the Ni signal is baseline resolved and therefore this wavelength can be utilized, while with the poorer resolution, which represents instruments of the early days of ICP-OES, the overlap by Si is so serious that the line cannot be used for Ni determination

Simultaneous array spectrometers have the flexibility of sequential spectrometers and the speed of simultaneous ones. The advice for method development with this group of instruments thus could be: the more lines that are checked initially, the better. However, the evaluation can get confusing if there are too many lines in a method. Every user has his personal preferences in this regard. The experience of the author shows that typically among five wavelengths taken at first, at least one line will be suitable, even in many difficult applications.

The fact that in simultaneous array spectrometers the detector can measure several wavelengths per element without time penalty, allows the user to take advantage of a number of wavelengths in the sensitivity range when starting method development with an unknown matrix. The unsuitable lines are then deleted one after the other. This process of method development is much faster than adding new wavelengths back into the method after a previous line has been found to be unusable, and one would have to start to record all the spectra all over again. This routine was suggested for conventional sequential instruments, particularly in the early years of ICP-OES [260]. For these types of instruments, the procedure of adding new lines makes sense, but for simultaneous array spectrometers the routine slows down method development unnecessarily. Sequential array spectrometers are a little slower than simultaneous array spectrometers, but clearly faster than classical sequential instruments. Therefore, the method development is usually faster if one checks several lines experimentally in one run.

In many cases, all lines are excluded but one. However, there are cases where two or more lines are deliberately kept in the method in order to back up the result of one wavelength with the other. This enhances trust in the analysis. In other cases, several wavelengths are used, because averaging the intensities of some lines of the same element with similar sensitivities improves the detection limit [261].

The most efficient way of line selection of course depends on one's particular method of operation and on the spectrometer and software used. The extensive possibilities of modern software (particularly for spectra storage and data reprocessing) allow the line selection to be carried out after the conclusion of the first measurement series. In extreme cases, the first routine series can be measured even though the method has not been developed to the end.

4.1.2.1 Generation of Spectra for the Estimation of the Background

In addition to interference by the matrix, argon emission lines (especially in the range from 300 to 600 nm, Fig. 78) may interfere with potential analyte lines [262].Furthermore, carbon and silicon lines are present in the spectrum of the plasma. Finally, rotation oscillations of molecules contribute to a distinctive molecular band spectrum. Molecular bands nearly always come in groups. The series of band emissions are distributed over a wide spectral range. Examples are shown in Figs. 79 and 80.

The most important species of molecular band spectra originate from OH radicals. These are found in the range between 280 and 330 nm. Oscillations of NO show at around 230 nm and N_2 molecular bands around 370 nm. (The intensity of the latter is greatly reduced by an extended outer tube on the torch that prevents entry of the nitrogen from the ambient air.) In a plasma into which organic solvents are introduced, oscillations of carbon compounds, especially C_2 oscillations in the range 450–650 nm, are observed.

The lines and molecular bands emitted from the solvent, the ambient air or the plasma itself can severely interfere with the analytical line [263]. The "advantage" of these interferences consists in the fact that their intensity is practically constant, and therefore they are easily recognized and corrected for. In the axially viewed plasma, there are more structures than in the radially viewed plasma. This is because of the higher sensitivity, which affects all components present but which one would prefer not to in this case. Here the advantage of the increased sensitivity becomes a disadvantage. When viewing the plasma axially, one also looks into cooler areas, which are hidden behind the induction coil when viewing radially. At the inlet end of the analyte channel there are molecules that have not yet been dissociated. Therefore, interference by molecular emission band structures is much more pronounced in axial viewing.

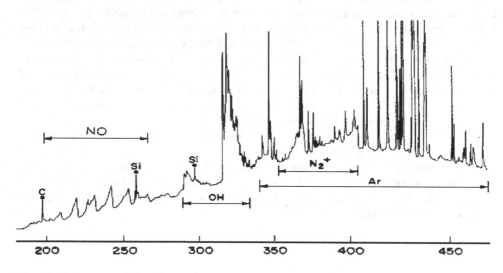

Fig. 78: Spectrum of the plasma between 175 and 475 nm. The horizontal double arrows indicate the wavelength range where certain molecular rotational bands (NO, OH and N_2) and argon emission lines are particularly frequent (source: PerkinElmer Instruments)

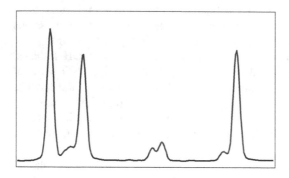

Fig. 79: Example of OH bands in the wavelength range of 309.193 to 309.497 nm

Sometimes a structure of the plasma such as an argon emission line or a molecular band will be misinterpreted as an emission line of the analyte. The spectrum of the blank solution will clarify this. If a signal is noticed at the place where the analytical line is expected, two interpretations are possible: it could be spectral interference, which would mean that the wavelength is unsuitable for the working range, or it could be contamination. In the latter case, the line would be usable, but certain steps in the analytical procedure should be reviewed. To gain a clear picture, the blank solution should be measured at least once more, preferably several times. If the signal has become less intense after a period of washing with the blank solution, then it is carryover of the analyte (memory-effect, Fig. 82, cf. also Sect. 2.4.2 "Nebulizer chamber"). Appropriate operating steps such as a prolonged rinse between solutions should be recorded as an additional result of this experiment and included in the method.

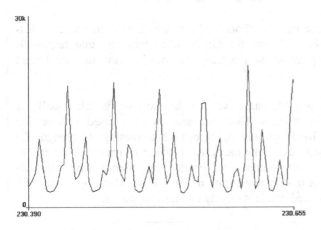

Fig. 80: Molecular bands in the wavelength range around 230 nm when aspirating organic solvents (in this case: kerosene). Because of the strong molecular bands, alternative wavelengths are preferable

If the spectra of the blank solution obtained in a series of runs match each other perfectly (Fig. 81), this could indicate either spectral interference or contamination of the blank solution. In order to find the actual reason, a few spectra of the blank solution should be taken again. The blank solution should be freshly prepared, taking all possible precautions to exclude contaminants. In unclear cases, it may be advisable to produce spectra of the pure solvent (typically water). Sometimes distinct differences arise if the solution is not prepared in glass containers (e.g. volumetric flasks), because the surface of the glass may absorb or release anlytes. Especially with the blank solution, extreme care should be taken: Because errors which go unrecognized here will sneak into the later measurements and inevitably lead to wrong numbers instead of the correct results [264]. This is even more true if the analyte concentration is near to the limit of detection.

✘   **Recognize peaks and distinguish them from structures**

There are cases in which the background has a structure, and it can be difficult to distinguish a peak from structured background. Two criteria can be consulted in order to make the distinction:

  (a) Regular structural characteristics
  (b) Intensity or intensity differences in various solutions.

Regarding (a): Structures are usually "lines" which are distributed more or less evenly over the spectral range (Figs. 79 and 80). On the other hand, an analyte peak is typically an isolated line whose peak position agrees well with its previously calibrated position.

Regarding (b): As a rule, a peak has a very high intensity, which usually is considerably higher than that of a structure. This statement should be used with caution because the background emitted by the plasma increases with increasing wavelength. Of course, the intensities of the structures on the background will also gain in intensity. In addition, in some applications (e.g. organic solvents) very intense structures are sometimes observed. If intensities in different solutions are compared, the intensity of a structure remains virtually unchanged, while it clearly changes for peaks.

✘   **What information can I obtain from the spectra of a blank solution?**

If one repeatedly measures the blank solution, the following questions can be answered:
  • Is there contamination?
  • Are carryover effects observed?
  • Do structures interfere with the determination of small concentrations?

For the cases mentioned, some example spectra are shown above in Figs. 81–83:

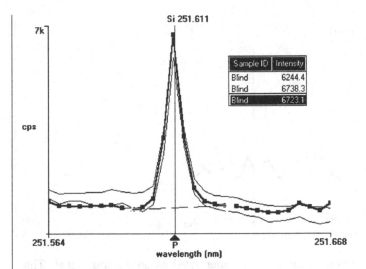

Fig. 81: If there is contamination in the blank, as is shown for the example of Si, the spectra remain practically unchanged. A good RSD is obtained for the repeats of the measurements

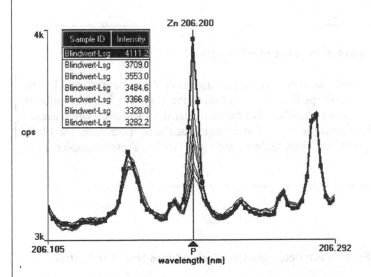

Fig. 82: Decreasing signals for repeated measurements of the blank solution indicate a memory-effect. One way to limit the effect of the carryover is to limit the amount of the element causing the memory-effect as much as possible by choosing a concentration as low as possible. Additionally, the rinse time between the solutions should be made as long as possible

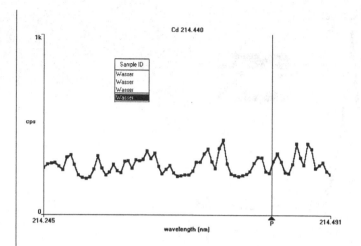

Fig. 83: A background structure should not be interpreted as an analyte signal. The spectrum of the blank solution is shown at the wavelength of Cd (214 nm). All spectra are congruent. Spectra of structures are furthermore characterized by a series of structures near the potential analyte peak

✗   **Beware of contamination in the blank solution!**

Particularly when working with an auto-sampler, measurements can be distorted by the sample capillary carrying over drops from one solution to the next. The blank solution is particularly at risk. In some cases, the blank is aspirated after the standard, which causes a contamination error right from the start of the routine analysis. Therefore, the blank solutions should be replaced at least before each run with an auto-sampler as a precaution.

4.1.2.2 Generation of Spectra for Detecting Interfering Lines from the Matrix

To check which elements and their corresponding lines interfere with the analytical line, wavelength tables should be inspected for potential interferents. Most wavelength tables are of limited use for one reason or another [265]. Therefore, after examining the wavelength tables, the spectra of all potential analytical lines should be recorded and critically evaluated. The first step is to generate the spectra of all calibration solutions to be used later in the analysis run.

ⓘ **Wavelength tables – not always as exact as they appear!**

Unfortunately, almost all wavelength tables used today in ICP emission spectrometry were compiled before inductively coupled plasma existed, e.g., to name the most important ones, the "MIT Wavelength Tables" [266] and the "NBS Tables" [267]. In principle, this should not matter: the emission line should not care whether it was recorded with a spark emission spectrometer or an inductively-coupled plasma. This is almost true. However, completely different excitation conditions exist in spark emission and plasma. Consequently, some lines emit strongly in a spark but weakly in the ICP, where they are visible only at high concentrations. Sometimes, important lines are missing in the tables. An example is the known aluminum line at 220.463 nm, which can considerably interfere with the determination of Pb.

Without exception, older compilations make use of data generated with a direct current plasma or spark emission even if "ICP" appears in the title [268]. Boumans states exactly in his compendium which factors were used to convert the data to the intensities that represent the conditions in an ICP [269]. The two-volume table "Line Coincidence Tables" published in 1980 was quite useful (unfortunately it is out of print), because it lists numerical values which show the influence of potential interferents on an analyte line as a function of resolution [270].

Winge, Peterson and Fassel (1979) were the first to publish a table with ICP emission lines that were actually generated in a plasma [271]. Unfortunately, the Fassel table contains only the strongest emission wavelengths ("prominent lines"). It is useful for the knowledge of which lines exist for the analyte. However, in most cases it cannot be used to forecast spectral interference. Important lines (e.g. B 182, Ca 315, K 766) are also missing.

The working group of Fassel published an atlas of ICP emission spectra in 1985 [272]. It gives a good idea of sensitivity and potential overlap.

In the same year, a table by Wohlers was published [273]. It contains weaker lines and thus supplements the original Fassel table. However, the rare earth elements are missing completely. These were added in 1989 by Boumans in a list of 1075 wavelengths of the rare earth elements and their interference on each other [274]. Here, the concept of quantifying the impact of an interference effect known from the "Line Coincidence Tables" was continued. In 1987, a collection of 6700 tungsten lines was published [275].

Another drawback to the tables is the fact that there are differences in the positions of the lines. The main problem is the calibration of the wavelength axis, a topic which has been under discussion for more than a decade by, among others, the National Institute of Standards and Technology. (At the time of printing of this book, this endeavor was still without result.) The major difficulty in the tables lies in the fact that only a small number of wavelengths were taken as reference points and the lines in between were interpolated. Depending on starting points, differences ranging from a few pm to a tenth of an nm can arise. This becomes especially obvious in tables that were compiled without comment from different sources. Here it can happen that single lines are listed several times with differences in the third or even second decimal digit (wavelengths in nm). One also finds similar differences when comparing different tables. An example of these contradictory entries is the position of the Zn line at 206 nm. In some tables the position is listed at 206.191 nm, while in another the position is given at 206.200 nm.

In the vacuum-UV range (<190 nm) the differences become particularly large: There the entries vary by up to around 100 pm. This is because of the difference between the refractive index in an inert gas and in that a vacuum. When using reference wavelengths outside the vacuum-UV and then extrapolating, the difference in the refractive index will show very clearly if some emission lines were determined in a vacuum and others in a nitrogen atmosphere. Even leaving this aside, in the nature of the methodology an extrapolation has a clearly larger uncertainty compared to an interpolation. Repeated use of the values in the tables without detailed information on the medium in which the wavelengths were originally measured could also be part of the cause of the confusion presently to be seen in some of the tables of ICP emission lines.

Also with the aim of giving exact wavelength positions, Schierle and Thorne in 1995 presented data generated by Fourier transform ICP-OES [276]. The quality of their data is excellent. Unfortunately, the data set contains only relatively few lines, so that its usefulness is restricted.

A common disadvantage of all work known to the author is the fact that important lines that occur naturally in the argon plasma (e.g. OH or NO bands) are missing. Since they interfere with many potential analytical lines, a list of these lines occurring naturally in the plasma would be extremely valuable.

The bottom line is that the wavelength tables are bits and pieces that do not always fit together well. There are many undocumented lines around, even more so since the sensitivity of ICP-OES was so greatly enhanced by axial viewing of the plasma. Therefore, it is highly likely that you will find a "new" line in the course of line selection in an accurate method development.

Obviously, an error must be avoided at all costs when calibrating. Therefore, the spectra of the multi-element calibration solution(s) have to be examined particularly carefully for potential spectral interference. In addition, the inclusion of spectra of single-element solutions in the evaluation is strongly recommended. A typical beginner's mistake is to select a wavelength which is just a little off the correct one. In order to identify the analytical line unambiguously, the user should know (i.e. record) the spectra of the pure analyte. Only the identity of the position (in the context of the positioning accuracy of the spectrometer used) can lead one to the adequately substantiated conclusion that the right element has actually been measured.

If there is a suspicion that there is a wavelength drift of the spectrometer, the single-element solution of the analyte should be measured repeatedly, one after the other. Through this procedure, one gets an idea of the wavelength stability [277]. If the wavelength positions are always found at the same spot, the stability can be assumed to be good enough.

If differences in the wavelength positions of the peaks for single-element or multi-element solutions and samples are observed in an instrument with good wavelength stability, then it must be concluded that the peaks are actually at different positions. This "shift" then actually points to two different emission lines, i.e. spectral interference.

If the wavelength stability is inadequate, the case becomes far more complex. In order to be certain that there is not spectral interference, all potential interfering lines must be excluded. The only safe way to do this is to record the spectra of all elements that emit spectra in ICP-OES in the range of the analytical line. For practical purposes, one frequently restricts the search to the elements which make up the main components of the matrix.

As well as spectra of standards, spectra of samples must of course be taken. These receive the most attention because the analysis of the samples is the main aim of the work. Ideally, the entire sample series is used for the method development. If the number of samples is small, this can easily be done. Furthermore, when using a spectrometer with hard- and software which collects spectral information at high speed, one can use the whole sample series without loss of time. If there is a very large sample series or if one is using a spectrometer where taking spectra takes a significant amount of time, one should choose a representative sample of the whole lot. This choice represents a great source of uncertainty in method development, since one can only rarely be certain which of the samples represent the sample series as a whole.

Beyond this, the author recommends the inclusion of atypical samples in the method development also. The advantage of this is that the emerging method is more robust, because it was developed using samples that show a greater spread. If it should turn out in the course of the method development that the atypical samples are not measurable with the same parameters as for the rest of the series, a special method must be developed for these.

4.1.2.3 Evaluating Spectra

The spectra recorded experimentally must be evaluated as the next step. The cases where spectral interference is clearly recognizable are straightforward. The easiest cases to identify are partial overlaps. Here a difference between the wavelength positions of analyte and interferent is clearly visible. A wavelength with such an overlap is definitely not usable for the analysis (at least with conventional processing). Unlike conventional processing, multivariate processing techniques are in many cases able to process the spectral information and calculate a correct result even for such lines which are clearly classified as unsuitable according the rules described. Recommendations concerning suitability of an analysis wavelength at spectral overlap therefore apply only to the classical processing techniques [278].

Sometimes small interfering peaks are overlooked or their impact is underestimated. A small signal can sometimes grow to a dominant signal and thus can cause severe interference in other samples of the series. Therefore, the user should identify all peaks near the analytical line. If it turns out that the emission signal arises from a main component, then in this case the line would be suitable to use, because the concentration of a main component cannot vary very much. However, if the peak originates from a trace element, then, as a precaution, a better analytical line should be sought, since the concentration of a trace can be subject to extremely strong variations and a clear overlap could suddenly result from a signal appearing to be insignificant.

The case where an isolated signal is observed is also trickier than one thinks. At first, this sounds like a paradox, but it is not. Of course, an isolated peak is the wish of every user of ICP-OES. Unfortunately, in such a case there is a (low) risk that an unrecognized direct overlap (line coincidence) can interfere the analysis. Therefore, the user must prove that there is no line coincidence. This direct overlap can be excluded safely, only if one takes spectra of all elements which emit lines in ICP emission. In practice, this requirement cannot be fulfilled in the routine laboratory, mostly because of time constraints. Therefore the test is frequently confined to the interferents in the matrix or where the wavelength tables suggest a direct overlap. It should be pointed out again that good spectrometer resolution minimizes the probability of a direct coincidence.

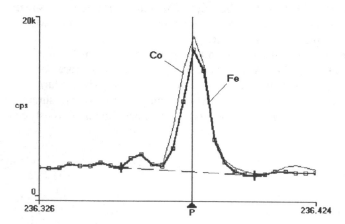

Fig. 84: If emission lines of analyte and interferent are found at the same position or if the wavelength difference is very small in comparison with the resolution, it becomes difficult to differentiate the emission lines from each other. In this example, the interference of Co at 236 nm by an Fe line is shown

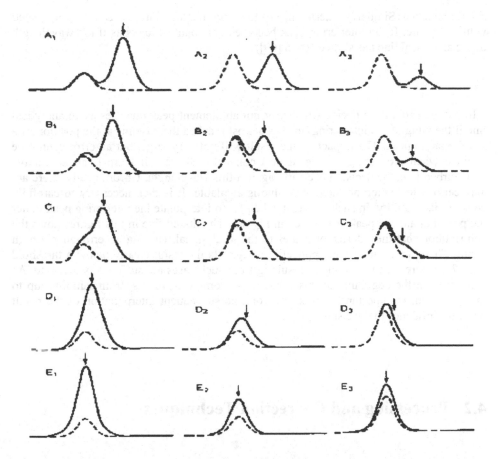

Fig. 85: In order to determine whether an interfering peak affects the result in a way that cannot be accepted, the peak is projected underneath the analyte peak. Here, the analyte peaks are highlighted with arrows. To the left of the analyte peak is an interfering peak. Its "wing" is interpolated to the right to demonstrate its influence on the calculated result. The peaks in the rows have different intensity ratios, in the columns the distances decrease. The distances expressed as multiples of the FWHH are: (A) 2; (B) 1.5; (C) 1; (D) 0.5 and (E) 0. The intensity ratios are indicated by indices. The intensity ratios of analyte peak/interfering peak are: (for 1) 1 : 3; (for 2) 1 : 1 and (for 3) 1 : 3. The intensity ratios combined with their distance in the top left part of the figure justify their choice as analytical line (particularly A1, A2 and B1). The bottom right part shows cases where intensity ratios and distances will clearly cause false results. Therefore, these lines must be excluded. It is particularly difficult to recognize the cases with direct line overlaps, as shown in the last row (E). Since a direct line coincidence can never be excluded although is becomes more improbable for a spectrometer with good resolution, the wavelength tables always should be consulted as an additional information source for line selection. If an interferent is recorded in the range of half a resolution, one should try to assess its potential influence on the analytical line. In order to be on the safe side, a single-element spectrum of this element in approximately the concentration expected in the samples

should be taken. Similarly, interference of the kind illustrated in row D has to be treated with great care. If the interfering peak becomes dominant, it looks as if the wavelength of the analytical line has shifted (from [69])

It is more difficult to decide whether or not an element peak qualifies as an analytical line if the wing of the interfering line has almost reached the baseline at the position of a small analyte peak. The impact on the result will not very large, but an error cannot be avoided when processing the signal in a conventional way. If there is an alternative undisturbed line, the decision is clearly against this wavelength. Unfortunately, there are applications for which no alternative line is available. It is then necessary to carefully consider its usability. In such a case, it is helpful to interpolate the interfering peak under the potential analyte peak as outlined in Fig. 85. The dotted line in this figure shows the contribution that the interferent makes to the total signal, causing an erroneously high result. There are a number of applications (e.g.: "Is the result clearly below a threshold value?") where the measurements resulting from such cases are sufficiently accurate. As mentioned at the beginning of this chapter, the complete process, from sampling up to interpretation, is important. In this situation, the subsequent interpretation of the result will thus influence the decision.

4.2   Processing and Correction Techniques

4.2.1   Signal Processing

The spectrum gained with help of the spectrometer is interpreted and used to calculate an intensity value for the signal maximum. Ideally, the analytical signal has a Gaussian distribution. For the description of this signal, the term "peak" has become accepted. The task of the signal processing technique consists of transforming the spectral information (the peak) into a numerical value (the intensity). If the spectrometer provides information about only one point, then this is automatically used for quantifying the signal (so called "on-peak" measurement), and further consideration of this topic is superfluous. However, there are a number of instruments (e.g. scanning sequential instruments or array spectrometers) where more spectral information is provided during the routine measurement. The intention of this additional information is to evaluate the signal during the analysis to minimize the risk of an erroneous measurement or false interpretation (e.g. to use the wing of an interfering peak as the signal).

If the spectrometer provides sufficient spectral information, then the peaks can be processed in different ways. First of all, only the techniques commonly used in most spectrometer software packages for processing ("classical processing") are treated in this chapter. A separate section is then dedicated to multivariate regression. For classical processing, two techniques are frequently used:

- Calculation of the maximum with a function that matches the Gaussian distribution at its maximum.
- Summation of the points, which represent the signal.

4.2.1.1 Calculation of the Peak Height

This method is designed to calculate the peak height from the signal even if the exact position of the maximum has not been measured. In this method of processing, a Gaussian distribution is presupposed for the analyte peak. At the maximum, a Gaussian function resembles a quadratic function. This fact is exploited for the calculation of the peak height, i.e., the points that characterize the peak in the upper part are put into a quadratic function. The maximum of the function is then calculated and equated to the peak maximum. Originally, this method was used for spectrometers whose wavelength stability was unsatisfactory. Using this dynamic method of peak processing, even a "drifting" peak can be processed with good precision. Because of this effect of dynamically "capturing" a potentially moving peak, this routine was originally named "peak search" . This method of processing has a slight advantage, even for spectrometers with very good wavelength stability, e.g. for processing intense signals when the highest possible accuracy is required, but this advantage is hardly significant in normal practice. Its precision is a little better than that of the alternative, which is area processing. The reproducibility improves, because the algorithm puts the emphasis on the signal height, giving slightly improved reproducibility in comparison to the mere addition of the raw data. However, for signals, which hardly differ from the base line, the noise pattern is much increased because of the emphasis on the height. This has the consequence that the intensities determined may turn out to be erroneously high. In addition, the precision and thus the limit of detection deteriorate in comparison to the summation of the measured points.

If the peak height processing technique is used, then the user generally can define the size of the window in which the peak maximum is expected. This window should be large enough to allow meaningful processing. At least three points on the signal are required to describe a quadratic function. More points yield a better fit for the calculated maximum. On the other hand, the window should not be too large, because otherwise there is the danger that interfering emission lines, where the wings reach into the window, are processed erroneously instead of the analytical line (Figs. 86 and 87 and Table 6). Peak height processing is utilized preferentially for isolated peaks where there is no interfering line in the immediate neighborhood.

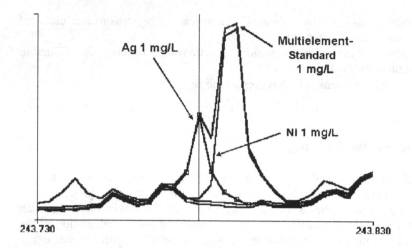

243.730   243.830

Fig. 86: Spectrum of Ag and Ni at 243 nm with a concentration of 1 mg/L each, in single-element and multi-element solutions respectively. For this spectral case, the results were calculated for height and area processing at different window widths

Table 6: Calculated concentrations of Ag at 243 nm in the presence of an interfering Ni line by the peak height method and using various window widths

Window width [pm]	Calculated concentration [expected = 1 mg/L]
3.7	0.95
11.0	0.95
18.3	2.09
25.7	2.19

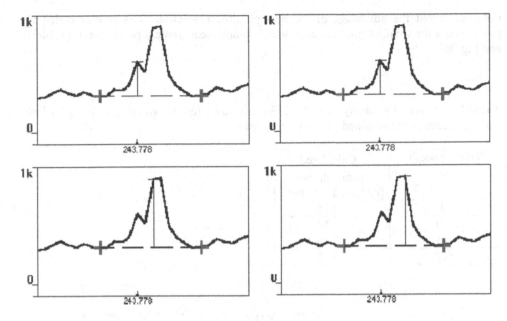

Fig. 87: Processing of the spectrum shown in Fig. 86 by the peak height method using four window widths, as shown in Table 6. The window widths increase in the rows from left to right. The vertical line shows where the peak maximum was "found" in the respective processing interval

4.2.1.2 Calculation of the Peak Area or of the Partial Peak Area

Another option to process the intensity is to sum the measurement points that describe the peak. This can also be described as an area integral. In this case, the area integral refers to the wavelength axis and not to the time axis, as it is common for other area processing techniques typically used in analytical chemistry. The area processing presupposes sufficiently good wavelength stability. It is always applied with advantage if there are interfering lines near to the signal to be processed. For small signals, area processing does not show the disadvantage of peak height processing, namely that the noise component is magnified. Consequently, in these cases it is also the preferred processing technique.

Area processing much resembles on-peak measurement. In some array spectrometers, the user can define how many measurement points are to be summed to calculate the intensity value for area processing. In general, the best signal/background ratio will be obtained by taking only one measurement point. (This corresponds exactly to on-peak measurement). Hence, single-point processing has a positive effect on the limit of detection. In addition, if there are spectral interfering lines nearby, limitation to only one point can minimize the impact of an interferent. Consequently, in analogy to the statement about the advantage of the best possible resolution, it is recommended to process only the smallest possible area, which is one measurement point (pixel) (Table 7 and Fig. 88).

Table 7: Calculated intensity of Ag at 243 nm in the presence of an interfering Ni line with peak area processing and various area sizes

Window width [pm]	Calculated concentration [expected = 1 mg/L]
3.7	0.95
11.0	1.14
18.3	2.10
25.7	3.06

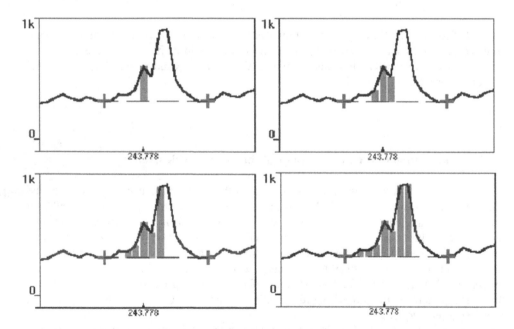

Fig. 88: Processing of the spectrum shown in Fig. 86 by area processing using four window widths, as listed in Table 6. The window widths increase in the rows from left to right. The gray boxes mark the partial surface areas used for the calculation. If too many points are chosen to calculate the area, then potentially interfering peaks and/or the background are included in the result, yielding worse accuracy and reproducibility (and consequently worse limits of detection)

Practical experience sometimes contradicts these theoretical statements. Area processing with only one point presupposes excellent wavelength stability, but if even minute fluctuations appear in the wavelength position, these will seriously affect the calculated intensities. Therefore, for reasons of good long-term stability, it can be beneficial to depart from the requirement to use the maximum (only one measurement point) and to include the adjacent points in the area calculation.

My instrument is sensitive to weather changes.

A rough weather forecast is based on the (barometric) air pressure. People predominantly notice the weather through their sensitivity to temperature. These two parameters are also important for the stability of ICP emission spectrometers

The main effect of temperature is on the angles of the grating, but also on other optical components by causing mechanical movements and twists of the optical chassis, changing the geometrical position at which the wavelength falls on the detector. To compensate for this influence, some spectrometers are equipped with a thermostat, which usually heats the optics to a predefined temperature. This temperature is typically set at a few degrees above the highest anticipated room temperature. If the room temperature rises above that, the (heating) thermostat cannot compensate for it. However, if the temperature were to be set even higher, this would mean a yet higher energy consumption and much more waste heat to be conducted away.

For this reason, ICP laboratories are often air-conditioned. This is done either to reduce the temperature maxima so that the thermostatting will work even in an environment where the temperature changes a lot or else to ensure an acceptable wavelength stability for instruments without a thermostat. (In the summer heat, this can be very welcome for the instrument operator as well as the spectrometer. Moreover, some computers tend to have higher failure rates at temperatures above 30 °C.)

Once the effects of temperature are under control, what remains is the influence of the ambient air pressure. This can affect the refractive index of the gas in the optics and hence can cause wavelength drift. This influence is observed mainly in spectrometers with superior resolution, a narrow exit slit and single point processing. Since it is not practicable to shield the spectrometer against the outer air pressure, a number of active wavelength correction mechanisms are used: The change in the position of a reference peak emitted in a natural ICP emission spectrum (e.g. carbon or argon lines) can serve as a correction factor. Further technical methods of correcting the wavelength position are the use of a mercury vapor lamp, which is periodically moved into the light path so that the position of an Hg line can be monitored, or else a neon spectrum is diffracted onto a part of an array detector not used analytically. The action to compensate for the wavelength change in the optics is to move the entry slit or the grating.

4.2.1.3 Calibration of the Peak Position

As a rule, the precise determination of the geometric position at which the peak has to be measured is closely linked to the peak processing algorithm. If the number of points to measure the peak is high enough to generate a faithful representation of the image or the measured spectrum is smoothed mathematically, the user will intuitively and correctly define the maximum as the peak position. It is more difficult if the number of the points that make up the spectrum is limited and shows a grid-type pattern, as is observed for the solid-state detectors used at present. In this case, it is advisable to calibrate the wavelength position at the peak center rather than at the highest point, which will not necessarily represent the original peak. The correct position for setting wavelength is clarified in Fig. 89.

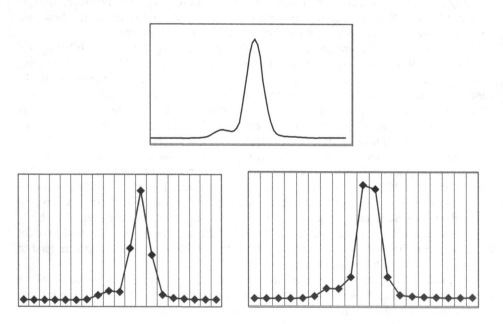

Fig. 89: Because of the relatively large pixel width in comparison with the FHHW, the image of the spectrum has a grid-type appearance on an array detector. In the figure, the "original" spectrum is displayed at the top, while at the bottom are shown two possible images of this peak produced by the detector from the original spectrum. If the peak maximum falls on the center of a pixel, a symmetrical peak results, as shown in the figure at bottom left. In the figure at bottom right, the peak maximum lies on the line separating two pixels, and the resulting image has a flat top. The original maximum is no longer recognizable. However, the exact position of the maximum in the wavelength axis lies in the peak center. This of course applies to all stages in between

4.2.2 Background Correction

The light emitted by the plasma contains numerous argon emission lines and a continuum that extends over the complete wavelength range used in ICP-OES (Fig. 78). The emission originates from the recombination of argon ions with electrons and the bremsstrahlung of the electrons. The intensity of the background increases with increasing wavelength for a coupling frequency of the RF generator of 40 MHz (Fig. 90). For a plasma driven by a 27 MHz generator 27.12 MHz, the background emission has a maximum at 450 nm [279].

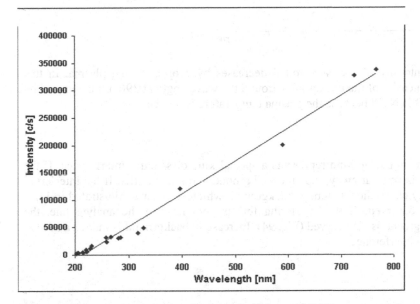

Fig. 90: The background of a plasma driven by a 40 MHz generator increases with the wavelength

Matrix elements from the sample can increase or decrease the background emission, so that its level must be measured and corrected for at every measurement. The smaller the analytical signal is in proportion to the background, the more important is this correction. The intensity of the background decreases if the sample introduced to the plasma causes it to cool down (Fig. 91).

The intensity of the peak base increases if a matrix element has a very intense emission line near to it. Since a wider base (a more clearly noticeable Voigt profile) is then observed, a very strong line even at a distance of some nanometers can be detected by an increase in the background emission. In addition, a number of elements show radiative recombination (of the ion with an electron) continuum, especially in the lower UV range (e.g. Al at 193–210 nm), which not only causes an increased background but also a heavily structured one [280].

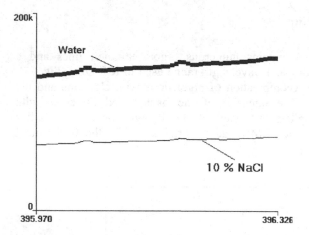

Fig. 91: The intensity of the background decreases by cooling of the plasma. In this example, the intensity of the background around the wavelength at 396 nm is reduced by a solution of 10 % NaCl because the plasma temperature is reduced

Thus, increase in background represents a special kind of spectral interference. If the interfering line is very far away, then the background shifts in parallel. If the interfering line is relatively close, then a sloping background, which is linear as illustrated in Figs. 92 and 93, is observed. If the interfering line is very near to the analyte line, the increased background is also curved (Fig. 94). Increase in background is indeed a special kind of spectral interference.

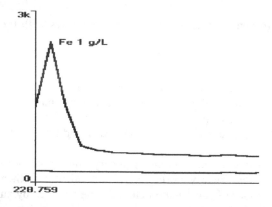

Fig. 92: Increase in background at 228 nm caused by introducing a 1 g/L Fe solution into the plasma (upper spectrum). The lower spectrum (horizontal line) represents the background emission of the plasma when introducing water

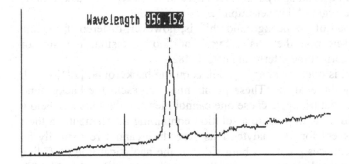

Fig. 93: Rising background in the region of the Al line at 396.152 nm caused by a neighboring Ca line (generated with a PerkinElmer Plasma 400)

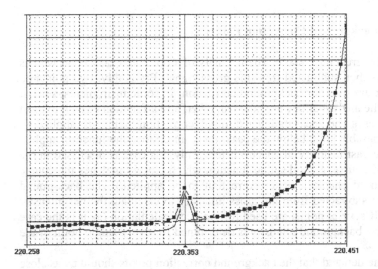

Fig. 94: The background of the Pb line at 220.353 nm is increased by an Al line (on the right outside the spectral window shown). Because of the proximity of the Al peak, the background is curved strongly (upper spectrum). For comparison, the spectrum of an aqueous solution of Pb is shown (lower spectrum) (source: Varian)

In the literature, structured background is listed under the topic background [281]. However, a structured background should be considered to be a spectral interference rather than a background, especially when the overlap with the analytical line is almost direct. This kind of background interference cannot be compensated by the usual method of background correction. Instead, because of the structured background, the wavelength should be regarded as unusable as an analytical line. Therefore, "structured background" by its nature belongs in the category "line selection" .

Reaction to the phenomenon of the background shift is, however, different from line selection. While for line selection there is a "yes" or "no" decision, the aim of background correction is to correct the interfering effect [282].

The background correction is done by choosing points on the background [283] which are always measured during the analysis. These points must represent the background below the line. This is somewhat tricky, because one cannot measure the intensity below the peak. The best one could do is to prepare a solution containing all elements in their respective concentrations except for the analyte. This procedure cannot realistically be performed. Therefore, one chooses points on both sides of the analytical line ("off-line correction") that characterize this background as well as possible, as outlined in Fig. 95. This demand is best fulfilled, if the background correction points are as near as possible to the analytical line.

4.2.2.1   Positions of Background Correction Points

As a rule, background correction is static. The intensity of a point is measured as well as the analyte signal. It is then subtracted from the gross signal. Preferably, two points on either side of the peak are used for background correction. In some instruments, several points on one side of the analytical line are taken together, and a function (which in most cases is linear) that uses the measured background correction points is applied to calculate the non-measurable point below the peak maximum. Since the function applied is a straight line in the vast majority of spectrometers, only this case is discussed in the following passage. (Of course, a good spline function which closely matches the course of the background would enable a better correction to be applied.) The background intensity determined is subtracted from the gross intensity to calculate the net intensity of the analyte signal. If several points are used for the peak calculation (as it is the case for area processing), the background intensity is determined for every single point of the peak.

In contradiction to the demand that the background correction points should be as close as possible, they should also not be too close to the analytical line because otherwise there will be a danger that the correction will lie within the wings of the analytical line. Of course, this should be avoided. For this reason, there is a "rule of thumb". The background correction points should be at least twice the distance of the FWHH away from any line ("twice FWHH rule"). This can be derived as follows. In the idealized state, the analytical line shows a Gaussian profile. For the Gaussian distribution, the "zero level" is reached theoretically only at infinity. For practical purposes, the wing becomes indistinguishable from the baseline at twice the FWHH (cf. also Fig. 96). However, this rule should not be overvalued, as the example in Fig. 95 illustrates. If the left background correction point were still further away, an erroneously low reading would result for some samples because some structures appear there.

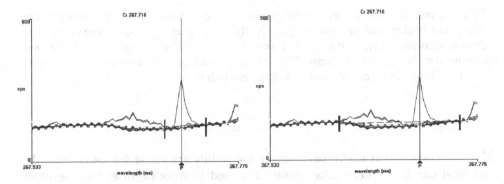

Fig. 95: The line calculated with the aid of the background correction points should be on the same level as the measured background in order to calculate a correct net intensity. In the left spectrum, a good position for the background correction points is shown, while in the right spectrum too much is subtracted. Particularly for small signals, this will make the result too low, or in the worst case will lead to a calculated "negative" concentration

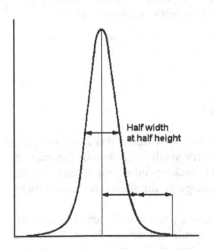

Fig. 96: At a distance of about twice the FWHH from a peak, the wing has the intensity of almost zero

To complete the picture, it must be made clear that the background correction points may not be positioned on any interfering lines or other structures. Therefore, it is recommended that, for setting background correction points, the scale of the spectra should be stretched so that the noise of the background emission will be clearly visible. Only by zooming in on the background, can small interfering peaks be recognized.

To minimize the risk of setting the points wrongly, some software packages can position the background correction dynamically. During every single measurement, a software algorithm checks the spectral situation of the background to find the best locations for the correction points. This dynamic background correction works quite accurately if the measurement window is large enough [284].

✂   **I suspect that background correction points are set so far away from the analytical line that other peaks between the analyte wavelength and background correction point will cause a false result.**

In contrast to the treatment of chromatographic peaks, the background correction points in ICP-OES are not start and stop points for the peak integration. In ICP-OES, the background correction points serve as points for the calculation of a function (a linear equation in most cases) that represent the real background as closely as possible. A vertical line drawn from the peak maximum (mathematically) onto this line is gauged to calculate the net intensity of the analyte signal. It is better to keep at some distance from the analytical line rather than to correct in the wings of peaks or to put the correction points on interfering peaks or their wings, which would result in a false correction.

4.2.2.2 Number the Background Correction Points

As a rule, two points are used for background correction. Obviously, this is required for a (linearly) sloping background (Fig. 93). Also, for very small intensities of the analyte signal, two points are recommended. With two points, background emission noise can be averaged. This gain in limits of detection is an advantage of simultaneous measurement with array detectors [291].

To correct a background that has shifted in parallel, one point is sufficient, particularly for a very intense signal. However, a second point would not do any harm. Since the background shift often results from a very intense line in the neighborhood of the analytical line, one can assume that a small share of the background shift always contains a slope. In selecting two points, one is on the safe side if the intensity of the interfering line should rise dramatically, causing the appearance of a more pronounced slope of the background.

In some cases, it can be advisable to correct a curved background only on the lower side. (An error cannot be avoided in this case, so the best choice would be to select another line.) In principle, a curved background cannot be corrected with a straight line. Therefore, the inevitable error can be only minimized. Experience shows that for a strongly curved background, the points should be set in very close proximity to the analytical line, if necessary by violating the twice FWHH rule. When using two background correction points, the calculated net intensity is too low; it is too high when using one point (on the side away from the interfering line) as Fig. 97 illustrates. Only an in-depth examination of an individual case will help to clarify where the error in the result is lowest and in which direction a discrepancy is most acceptable.

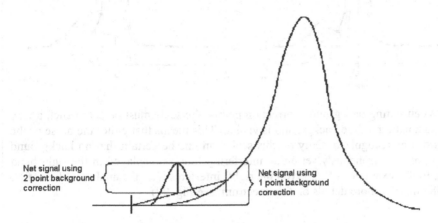

Fig. 97: Options for setting background correction points for a curved background. If a linear equation is used for background correction, the best fit will be often obtained by using only one background correction point on the side away from the interfering line

Sometimes (more often than one would like), there is no room in a spectrum for two interference-free background correction points because there are so many interfering peaks. In these cases, it is better to use only one background correction point rather than run the risk of putting one point on an interfering line or its wing.

As a final consideration, the mode of measuring the background correction points in spectrometers (sequential or simultaneous instruments) is an important issue, because sometimes an additional measurement must be made. Each additional background correction point takes time, and the extent of the increase in the duration of the analysis, must be assessed in each individual situation. In many cases, the additional time requirement is negligible. Although the maxim "time is money" may sometimes be justified, one should not run the risk of even larger eventual costs due to false results. Therefore, the same care as was invested in the line selection should be used for setting the background correction points.

🔧   **What can go wrong when setting background correction points?**

Figure 95 gave an example of the correct placement of the background correction points, and further examples are given below.

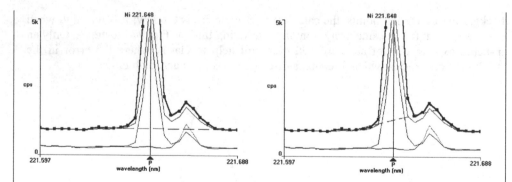

Fig. 98: When setting background correction points, the scale must be set in such a way that the exact pattern of the background is visible. This means that either the noise or the structures can be recognized. Only on this scale can one be certain that no background correction point is erroneously set on an interfering line, as displayed in the right hand spectrum. In this example, a Si line (on the right) interferes with the analytical line of Ni at 221.648 nm. (For more details on this spectrum, see Fig. 99)

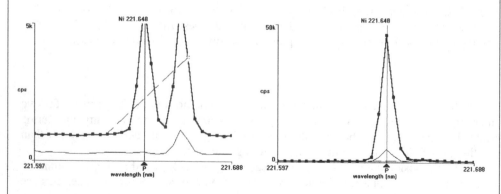

Fig. 99: When setting the background correction points, nearby peaks should be identified. If the concentration increases greatly, then a longer distance is preferred to prevent the background correction from being positioned on (the wing of) the interfering line. In this example of the determination of Ni at 221.648 nm, the concentration of the interferent Si is very variable. Specifically in the case of Si, there is a risk of contamination during sample pretreatment: Therefore, special caution is advisable here since the Si concentrations can sometimes be unexpectedly high in a sample solution. A significantly low reading then can result by correction in the wing of this interfering peak. This example is included to emphasize the point that background correction points should not be set too close to any peak (in this case in the wing of the interfering line). In addition, if the scaling is chosen unfavorably, as in the right hand spectrum, the interfering line can go unnoticed

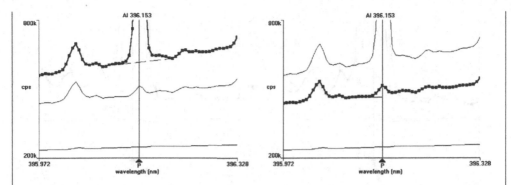

Fig. 100: If, for safety's sake, two background correction points are set on either side of the peak, as shown in the left hand spectrum, a potential slope is taken into account in each case. In this figure, the example is given of an Al determination at 396.153 nm in presence of high Ca contents. In addition, it is important here to use the correct scaling, as otherwise the slope might be overlooked. Consequently, a false correction could result with only one background correction point, as illustrated in the right hand spectrum. Especially for small signals, this can constitute a sizeable amount of the analyte signal

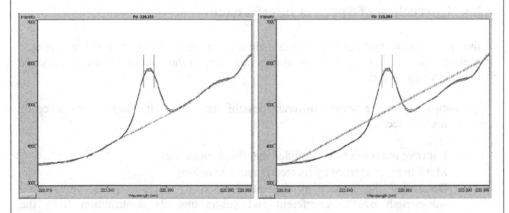

Fig. 101: The background correction points should be as near as possible to the analyte peak in the case of a curved background. An overcorrection (leading to an erroneously low result) is performed if the background correction points are set too far away from the analytical line. The example shows a Pb determination at 220.353 nm in the presence of high amounts of Al (source: Varian)

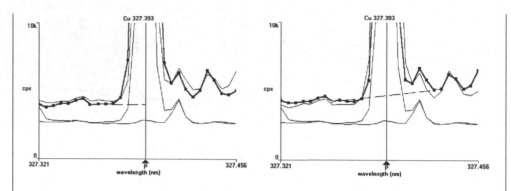

Fig. 102: If there is no space for a background correction point on one side of the spectrum, it makes sense to set only one point to avoid the risk of setting a background correction point positioned on a peak or its wing. The example shows the Cu line at 327.393 nm

4.2.3   Correction of Spectral Interference

In the course of method development, a main concern is to avoid spectral interferences by skillful line selection. This is not always possible, so the next best attempt would be to correct the interference.

In principle, there are several different possibilities for minimizing the influence of spectral interference:

- Improve the resolution (provided that this is possible)
- Mask the interference by (perfect) matrix matching
- Determine the impact of the interference by measuring an undisturbed wavelength of the interferent and subtracting its contribution from the interfering analyte line (inter-element correction)
- Apply a correction technique by multivariate regression.

In order to deal with spectral interference, the first priority is to opt for the best possible resolution. Some spectrometers offer the possibility to set different resolutions. In this case, the best possible resolution should always be chosen.

As its name suggests, matrix matching is designed to compensate for non-spectral interference. However, in individual cases it can also successfully be used in dealing with spectral interference. This is successful in cases, where the interferent has the same concentration in all solutions. An example would be iron in a batch of steel samples, or in cases where the intensity of the interferent does not change much, as in the molecular bands of the natural argon plasma emission spectrum.

4.2.3.1 Inter-element Correction

The oldest technique for correcting spectral interference is inter-element correction (IEC) [285]. This was developed in a time in which spectrometers usually had a resolution of 30 or even 50 pm. At this resolution, the peaks are extremely wide, and there are frequent direct overlaps. IEC was created for these direct overlaps. Since analytical and interfering lines are at identical position, a further wavelength of the interferent, which itself should be interference free, is needed. As part of the method development, the ratio of the two emission lines of the interferent is determined with single-element solutions. The ratio obtained is then used to determine and subtract the contribution of the interferent from the total signal intensity (of the interfered analytical line plus the intensity of the interfering line).

The analyte signal corrected by the impact of the interference is calculated as follows:

$$I_{Anal1,corr} = I_{\lambda1,total} - I_{Interf,\lambda2} \cdot m_{IEC}$$

The IEC factor m_{IEC} is first determined as:

$$m_{IEC} = I_{Interf,\lambda1} / I_{Interf,\lambda2}$$

where $I_{Anal1,corr}$ is the intensity of the corrected analyte signal at the wavelength λ_1, $I_{\lambda1,total}$ is the total signal (the sum of analyte and interferent signal) at the wavelength of the analyte λ_1, $I_{Interf,\lambda2}$ is the intensity for the interferent at the alternative wavelength λ_2, m_{IEC} is the ratio of the intensities of the interferent at wavelength λ_1 divided by that at the wavelength λ_2.

Partial overlaps can be corrected only to a limited extent by inter-element correction. These are difficult to correct because only the wing of the interferent peak is included. It is obvious that a measurement in the wing is not very reliable. Since inter-element correction is an additive correction, the analytical error always becomes greater in accordance with the law of error propagation. This results in worse reproducibility and higher limits of detection.

The IEC factor is dependent on excitation conditions. This is particularly true if one interferent line is an atomic transition, while the other is an ionic transition. If the plasma temperature drops by introducing a high matrix concentration, then the intensity of an atomic emission line can increase and the intensity of an ionic emission line can decrease. The parameters that determine the temperature of the plasma (RF power, gas flows) must therefore be absolutely constant. Otherwise, a false correction can result. Since the nebulizer gas flow has an essential influence on the sensitivity of a line, the state of the nebulizer is also extremely important for a correct analysis with IEC. Often, even in routine analysis, the IEC factors are determined immediately before the beginning of an analysis run. This means more work for the user, not only in the method development phase, but also in the regular routine.

For the IEC method development, all spectra of all the analytes as described in the general part are taken and evaluated. Then the processing parameters and background correction points are defined. After the identification of the interferent, an undisturbed wavelength must found for this element. Of course, background correction points have to be set for this peak also.

For the determination of the IEC factor, a pure single-element solution of the interferent is aspirated. Contamination will inevitably result in wrong IEC factors, and therefore everything must be done to exclude it. In a multi-element analysis, potentially mutual interferences must be taken into account. The determination of IEC factors with cross interferences, sometimes in the format of table, can become a quite arduous task [286]. Normally, the IEC factors are calculated by the spectrometer software.

4.2.3.2 Correction using Multivariate Regression

With increasing improvement of the resolution, apparently direct coincidences often change into partial overlaps. In order to have an optimal correction technique for partial overlaps, a chemometric technique was developed in 1990 (Kalman filtering) [287, 288]. This uses a multivariate regression calculation to correct partial overlaps correctly and efficiently [289]. It can be obtained as a post-analysis software package as shareware [290]. A similar procedure was introduced in 1993 for the first time in the software of a commercial spectrometer [291, 292], where it was called "Multi-component Spectral Fitting" (MSF). Another manufacturer of ICP emission spectrometers has used a similar routine since 1998 called the "Fast Automated Curve-fitting Technique" (FACT) [293].

The application of multivariate regression techniques is based on the fact that the light emitted by the plasma is made up by the addition of the single components to form the measured spectrum. Components in this sense are the single-element spectra and the background spectrum of the plasma. Since the background spectrum is an inherent part of all spectra, all single-element spectra must be corrected by subtracting the background spectrum, so that a "net" model spectrum results. The technique basically rescales all net spectra plus the background spectrum until a good match to the measured spectrum of the solution is obtained. One also could describe this routine as "arithmetical matrix matching".

Multi-component spectral fitting (MSF) was the first chemometric application to be implemented in the software of a commercial spectrometer. Therefore, MSF is described here as an example of the multivariate techniques. The mathematical concept describes the spectra of the solutions as vectors and the model spectra as rows of a matrix. By matrix transformation, the scaling factors can be calculated using the equation:

$$Y = X \beta + \varepsilon$$

where

Y	is the vector of the unknown spectrum ($n \times 1$)
X	is the vector of the model spectra ($n \times \rho$)
β	is the vector of the scaling factor ($\rho \times 1$)
ε	is the vector of the residuals ($n \times 1$)
n	is the number of points per spectrum
ρ	is the number of components.

For this equation, the residual error is minimized.

Multivariate regression processing takes place as follows. The measured spectra of the sample solutions serve as a basis to calculate the different scaling factors from *a priori* knowledge of all single-element model spectra involved and the spectrum of the plasma. The concentration of the analyte is calculated using the scaling factors rather than the intensities as normally done.

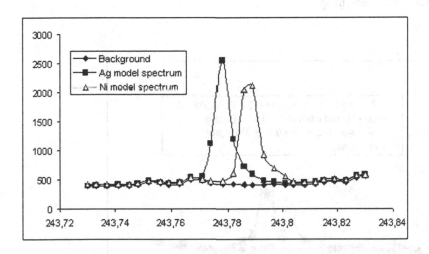

Fig. 103: Single-element spectra of the components are taken as part of the method development for multivariate regression techniques

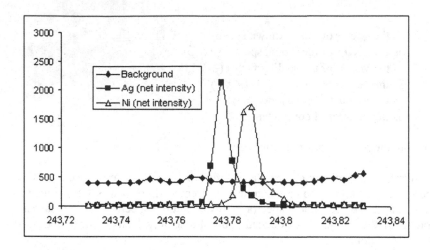

Fig. 104: The background spectrum is subtracted from the single-element model spectra to generate net single-element spectra

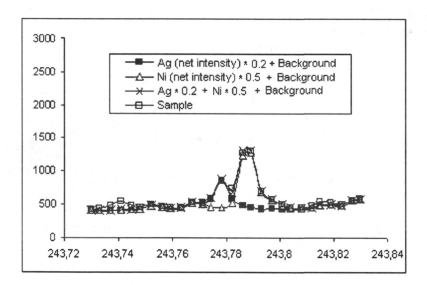

Fig. 105: The net single-element spectra are rescaled until the spectrum composed of the model spectra agree as well as possible with the measured spectrum. The scaling factors so obtained are used for the quantification

This method of calculation has a unique mathematical solution only if different structural characteristics (e.g. wavelength difference or further emission lines in the spectral range) exist. The quality of the correction as a function of the distance is shown in Fig. 106. From the diagram it again becomes clear that this processing technique is designed only for partial line overlaps (or complex spectra). In contrast to classical processing, it is true to say that the more complex a spectrum and the more unambiguous the multivariate regression processing, the better is the result. Since multivariate regression uses more spectral information than classical processing techniques, the statistical error decreases, so the reproducibility and thus the limits of detection improve [294].

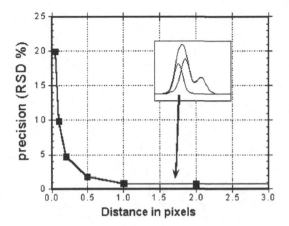

Fig. 106: The quality of spectral interference correction with multivariate regression techniques improves with increasing distance between the lines (source: Varian)

For multivariate regression method development, the spectra of all components in the spectral range (typically 0.1 to 0.4 nm) must be taken into account. It is important that these are pure substances, since otherwise erroneous results can be caused. As for classical method development, the spectra of the blank solution and the single-element solutions of analyte and interferent(s) are necessary, but the concentration used for the model spectra is not crucial. The lower limit for the concentration is the requirement that the spectrum should not be dominated by the background noise. The upper limit is that no self-absorption should occur, as this will lead to wider peaks as illustrated in Fig. 107. If these widened peaks are used as a model, good agreement between the model and the real spectrum cannot be obtained.

The method development for multivariate techniques becomes slightly more prolonged because the generation of single-element spectra of all the components is required. For classical method development, the generation of the spectra of single-element solutions is only recommended, but for this multivariate technique it is mandatory. The purity demands are also higher in certain regards. Only if the model spectra show no influence of other elements can they be used successfully for multivariate regression processing. However, this requirement on the absence of contamination only affects the elements that have emission lines in the wavelength range observed.

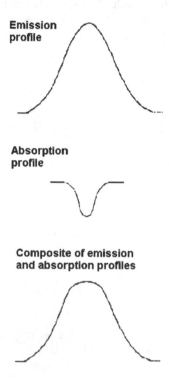

Fig. 107: In the hotter regions of the plasma, radiation is emitted (top peak). This can be absorbed again in the cooler zone of the plasma if the number of the particles is so large that the probability for resonant absorption exists (center peak). This occurs mainly in the cooler zones of the plasma. At lower temperatures, the absorption signal is narrower than the emission signal (because of the Doppler effect). The superimposition of the two profiles leads to a very wide peak which looks as if its top is missing (bottom peak)

As a compensation, method development for multivariate processing is simplified by the fact that spectral interference, as a rule, does not lead to the exclusion a line. Furthermore, it is significantly simplified by the fact that no background correction points need be set (in contrast to classic method development). Especially when there are interfering lines near the analytical line, the optimal choice of the background correction points is sometimes quite complicated. In the case of the multivariate processing techniques, all that needs to be done is to assign which background spectra are of analytes and which are of interferents (cf. Fig. 108). In many cases, the interference can be corrected by a multivariate regression technique even after the analysis has been completed.

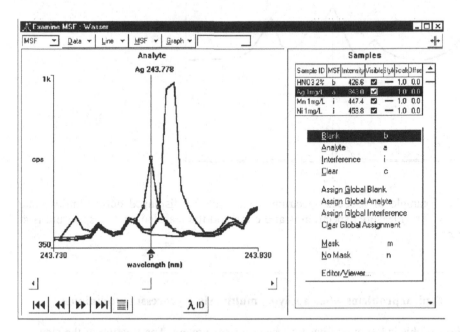

Fig. 108: Spectra used for the MSF method development. The example is given for Ag at 243 nm. The model spectra are marked with letters: The background spectrum is marked "b", the analytes "a" and the interferent(s) "i" in the legend on the right of the figure (screen from the PerkinElmer ICP WinLab Software)

For the successful application of multivariate processing techniques, it is an essential precondition that the wavelength positions should not change between the recording of the model spectra and the analysis. All means to assure wavelength stability must be utilized.

Since the signal form changes to a negligible extent according to plasma temperature, this processing technique is largely independent of the excitation conditions. This means that model spectra, which were taken for example at a low plasma power, can be used for an analysis with a high plasma power.

During analysis with multivariate processing, the model spectra are fitted to the measured spectrum during the measurement, this being the case both for the calibration solutions and for the sample solutions.

The corrected analytes spectra can be shown along with the measured spectra during the analysis in some software packages. This serves as a plausibility control. Figure 109 shows an example of such a screen display with such a measured spectrum and rescaled analyte peak.

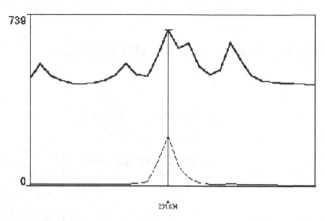

Fig. 109: Example of a check spectrum, which can be displayed during multivariate processing. The measured spectrum is at the top and the rescaled analyte spectrum is at the bottom

---

⚒   **Typical problems when applying multivariate processing techniques**

Impatience. This is the most frequent reason for carryovers. The fraction of the element carried over is interpreted as a component part of the spectrum of the following element during the model spectra generation phase. Consequently, it will be rescaled with it during the analysis. This causes the concentration calculated for the element affected by the carryover to be too low, and in the worst case a negative concentration will be computed.

To prevent carryovers, a blank solution should be aspirated between the solutions used for the model spectra. Only if one has convinced oneself that the background level has again been reached, may the next single-element solution be aspirated.

Contamination. The symptoms are similar and the consequences are the same (low results) as for carryover. The only remedy is to obtain or produce purer solutions.

Interferent not identified. All components are "twisted" to make the sum of the rescaled model spectra agree with the measured spectra. If one component is missing, the other ones will be used to make up for it. Consequently, a false scaling factor results for them. The remedy is to identify all elements appearing in the spectral window and generate model spectra for them. Alternatively, some software packages allow a part of the spectral range to be ignored for multivariate regression. This works acceptably only if the distance between analyte and interferent lines is large enough. (There may be no recognizable overlap even from the peak wings.)

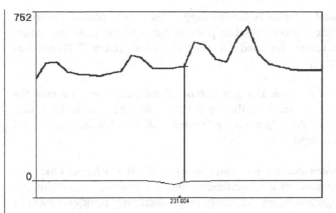

Fig. 110: A negative control peak can have various causes:
- An interfering peak was not taken into account.
- The solution(s) for the model spectra were not pure enough.

Concentration of the model solution is much too low. The spectrum of an element at a concentration close to the detection limit contains a strong noise component that becomes "scaled up". The consequence may be a too high or too low reading. Whenever the concentration of the analyte in the model solution is much lower than that in the sample solutions, structural characteristics, e.g. peaks arising from contamination, are rescaled with the same factor. The best results are obtained if the concentration used for the model spectra is higher (up to one or two orders of magnitude) than the anticipated concentration in the solutions to be measured.

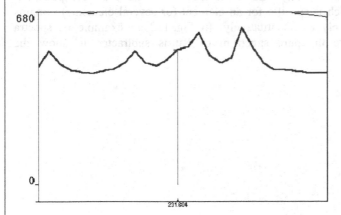

Fig. 111: The control peak is almost invisible. One can merely guess at its presence at the upper edge of the display window. This is a sign that the concentration in the model solution was too low

The concentration of the model solution is far too high. This only becomes a problem if the non-linear range is reached. Here self-absorption occurs and the peak maximum appears flatter. Such a peak cannot be used for rescaling of a "normal" (Gaussian-shaped) peak (in the linear range).

Wavelength drift has occurred between the generation of the model spectra and the analysis. A clear sign of this is an "oscillation pattern" for the check spectrum. Calculated results make no sense. As a "preventive" remedy, all possible measures must be taken to guarantee the wavelength stability.

The intensity difference between analyte and interferent peaks is too large. Once the analyte peak disappears in the noise of the interferent wing, the limits of this technique are reached. This can be recognized by an extremely large interfering peak and by very poor reproducibility of the computed result.

The wavelength difference is too low. Multivariate techniques transform spectra into rows of a matrix. The matrix transformation yields a clear scaling vector only if the structural characteristics can be clearly differentiated. This means that a coincidence or an almost direct overlap will not yield an unambiguous mathematical solution. The number calculated is a chance product and is subject to great statistical variation (extremely high RSD).

The technique of "spectra subtraction" can be seen as virtually a conceptual precursor to multivariate processing. In this technique, the complete spectrum (or that part in the wavelength range around the analytical line defined by the user) is subtracted from another. Typically, a structured background or the spectrum of the main component is subtracted. A spectrum which is specific for an element (or several elements) remains and can be used qualitatively or quantitatively. In Fig. 112 an example of spectra subtraction is shown. Here the pure spectrum of Ru is subtracted to show the contaminants in the Ru.

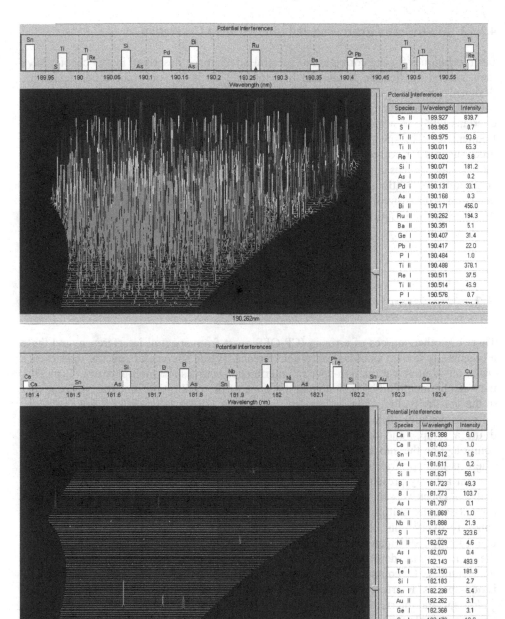

Fig. 112: Subtraction of one spectrum from another. At the top, the spectrum over the complete wavelength range for a Ru sample, which is slightly contaminated, is shown. From this the spectrum of pure Ru (not shown) is subtracted. The remaining spectrum is shown at the bottom. Here the vertical lines indicate the element lines that are present in the sample. They were identified as Ca, Mg, and S (source: Varian)

4.2.3.3 Multivariate Regression and Inter-element Correction

Both techniques for the correction of spectral interference described in this chapter have their special application fields and limitations. IEC is the only correction technique designed for direct line overlap and hence is the only one applicable in this case. Multivariate processing techniques on the other hand are only successful for partial overlaps [295]. In a modern spectrometer with good resolution, this is by far the most frequent form of spectral interference. Particularly complex spectra can be used with great success for correction with multivariate regression. An indisputable advantage of multivariate processing techniques is their independence from changing excitation conditions. In addition, multivariate techniques offer better reproducibility and limits of detection. IEC involves a significantly higher workload during both method development and routine use.

4.3 Non-spectral Interference

There are other sources of error in ICP-OES apart from spectral interference, and these are generally known as non-spectral interference. While sources of spectral interference are of an additive nature, non-spectral interference sources are multiplicative. Typically, they can be identified by a change in the slope of the calibration function (respectively of the sensitivity, Fig. 113).

Experimentally, they can be recognized in several ways. The simplest approach is to dilute a sample and determine the concentration in the diluted and in the undiluted solution. For the subsequent calculation, the undiluted sample serves as a reference. There is non-spectral interference if the amount of analyte determined divided by the amount actually present (the recovery rate) differs significantly from 100 %, provided that this check was made in the linear range. As a rule, the amount determined in the diluted sample multiplied by the dilution factor is more than 100 %. Alternatively, the spike recovery can be checked. Here the analyte is added in approximately the concentration anticipated in the sample. As a rule, the recovery will be too low; hence, the term "signal depression" is used to describe this characteristic of non-spectral interferences. Of course, in this case also the linear range should not be exceeded. There is non-spectral interference if the recovery rate differs significantly from 100 %.

Non-spectral interference can occur because of changes in the

- Sample transport (e. g. worse sample transport by the pump)
- Nebulizer properties
- Nebulizer chamber aerodynamics
- Excitation conditions in the plasma.

Non-spectral interference is caused by changes in the physical properties of the sample (particularly: viscosity, density, and surface tension), change in the rate of mass transfer into the plasma, temperature change at a constant RF power, or a change in the number of electrons in the plasma [296]. A higher RF power can reduce the impact of excitation interference [297].

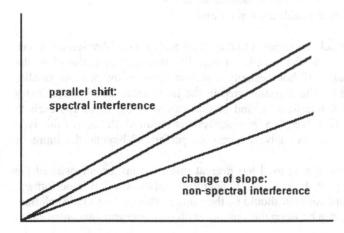

Fig. 113: Effects of spectral and non-spectral interference on the slope of the calibration function. When the sample solution is spiked with analyte, a parallel shift of the calibration function results for a spectral interference. The shift corresponds to the amount of analyte added. In the case of non-spectral interference, less than 100% of the amount added is determined in the analysis and the slope of the calibration functions changes significantly.

4.3.1 Correction of Non-spectral Interference

If non-spectral interference is encountered, a number of measures can be taken to avoid or correct the effects. These corrective actions are listed in the order of their effectiveness and practicability (in the author's experience):

1. Matrix matching
2. Use of an internal standard
3. Calibration by analyte addition
4. Addition of surfactants
5. "Ionization buffer".

4.3.1.1 Matrix matching

The most effective and most common option to avoid the impact of non-spectral interference is matrix matching [298]. This can be accomplished in two ways:
- Match the standard solutions to the sample solutions, or
- Match the samples to the calibration solutions.

   If the method used is to match the samples to (typically aqueous) calibration solutions, this usually means a dilution of the samples. Logically, this step is included in the sample preparation procedure. If homogeneity considerations allow this, a smaller sample mass could be used for the digestion step in the first place. Dilution eliminates many effects of non-spectral interference, and to that extent is the ideal approach to overcome this interference. However, if a trace analysis is required, the use of this type of matrix matching is restricted, as analysis cannot be performed beyond the limits of detection.

   The type and concentration of acid used has a great influence on the intensity of the analyte peaks (acid effect) [299, 300, 301, 302]. Therefore, when dilution is performed, the acid matrix in the standard solution should be the same as that in the sample solution. Also, it is advantageous to rinse between the samples with this type and concentration of acid [303].

   For optimal matrix matching, not only should the type and concentration of the acid be matched, but also the other ingredients such as digestion reagents and the main components of the matrix should be in roughly the concentrations present in the sample. One great risk of matrix matching, which exists particularly in the range of trace analysis, is that contamination could be introduced into the standard solution(s) , which can cause a faulty calibration. Therefore, particularly those elements which are added in the higher concentrations should be of the highest purity.

   Thus, it becomes obvious that for matrix matching, in some cases, a screening analysis may be useful to give a formula for a well-matched standard solution. In a number of applications, a standard reference material or a well-characterized sample is used for calibration. This is a kind of near perfect matrix matching. However, this option is regarded as very controversial by some workers, e.g. environmental analysis groups, while in others the method is well established, for example in steel analysis.

4.3.1.2 Internal Standard

If the samples are too different in composition and a dilution is not possible, the next best and most frequent option would be the use of an internal standard. The concept of the internal standard is based on the idea that the effect of the non-spectral interference is determined and then used as the correction factor [304]. An element is added to all the solutions during the sample preparation step in the same concentration in each case. The change in the intensity of this internal standard then serves as a correction factor for all other analytes.

The correction works successfully if the interference affects all elements or analytical lines in the same way. However, this is not the case for excitation interferences, which are based on a change in the excitation temperature in the plasma. In these, each line reacts differently according to its norm temperature. This circumstance is often either ignored completely or inadequately taken into account. Occasionally it is recommended that atomic emission lines should be used as internal standards for the correction of atomic emission lines of the analytes or that ionic emission lines should be used to correct for ionic emission lines [305]. Since the circumstances of the excitation of the emission lines in the plasma are fundamentally more complex, the uncritical application of this strategy of the internal standard does not always prove successful. In extreme individual cases, even a false correction can be carried out, as illustrated in Fig. 114. Rather it is necessary that the wavelengths of analyte and internal standard should have comparable excitation energies in order to obtain a correction as good as possible [306, 307]. Therefore, the spectrometer software ideally should enable an individual internal standard element wavelength to be assigned to every analyte wavelength.

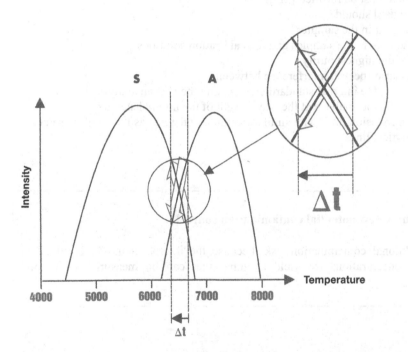

Fig. 114: When using an internal standard, the norm temperatures of the line pair (S for internal standard, A for analyte) should match as closely as possible. If there are very large differences, as illustrated in this figure, a false correction can result because of a temperature drop, as typically occurs when a sample with high dissolved matrix concentrations is introduced into the plasma. The intensity of the analyte decreases with decreasing temperatures, while the intensity of the internal standard increases.

When using an internal standard, one should work in the linear range, since in the non-linear range a good correction is extremely unlikely due to fact that the curvatures of the different lines are congruent. If this cannot be avoided, post-analysis data treatment using external programs is recommended [308].

Because the effects of non-spectral interference are generally more complex, a good correction sometimes cannot be obtained with only one strategy. There are a number of suggested methods of extending or modifying the concept of the internal standard in order to achieve a good correction [309, 310]. However, these procedures are not integrated into the software packages of commercial instruments and therefore are not normally suitable for routine analysis. A combination of matrix matching and internal standard gives a very good correction [311].

The use of an internal standard can yield a significant improvement in reproducibility and accuracy if the wavelengths of analytes and internal standards are measured truly simultaneously, with an array detector [312, 313, 314].

Before using an internal standard, consideration should be given to the following prerequisites, which must be fulfilled [315]:

The internal standard should

- be absent in the samples
- be soluble in the samples and the calibration solutions
- be of the highest purity
- not cause spectral interference between
    - the internal standard element and the chosen analytical lines
    - the analytes at the wavelength of the internal standard
- have emission lines with similar excitation energies as those of the selected analytical lines.

✘   **Caution: A new potential contamination source!**

There is an additional contamination risk, because the internal standard is added in relatively high concentrations (to yield a signal that can be measured with good reproducibility).

The practical procedure is to add the identical concentration of an element to all solutions (blank, standard(s) and samples). The concentration added should yield an intensity that can be measured with good reproducibility. Normally, this corresponds to a value of at least 100 times the limit of detection of the analytical line, to obtain a good quality corrected result. The instrument software normally calculates the ratios of the intensities of the analytes to the intensity of the internal standard. This ratio, rather than the intensity, is then used for the quantification.

✂ **Caution! The handling of an internal standard varies from instrument to instrument!**

For some instruments or their associated software packages, the strategy of the internal standard may differ considerably from that described here. Therefore, the user should find out which options and features are provided in the instrument software. Sometimes, no internal standard is added to the blank solution. In other cases, several internal standard elements can be added, or several lines of the same element (or even of different elements) can be used for the correction.

4.3.1.3 Calibration with Analyte Addition (Standard Addition)

One option to compensate for non-spectral interference is the application of the technique of analyte addition   (synonymous expression for standard addition) for calibration. This allows an optimal correction of the non-spectral interference. The procedure is very time consuming for a multi-element analysis. As a rule, other ways are preferred to compensate for non-spectral interferences. However, there are cases where the matrix effects are unclear or only a few samples are to be analyzed. Then the technique of analyte addition is faster than matrix matching or the search for an internal standard.

When using the method of analyte addition, the slope m of the calibration function is determined afresh for every sample. Practically, a known quantity of the element to be measured (spike) is added to the sample. The slope m can then be determined from the difference between the emission signals (ΔI) divided by the added final concentration (Δc), as shown in Fig. 115. Next, the intercept must be determined. This is done by measuring a blank solution or setting the intensity to zero. The concentration in the sample can then be determined.

Since the technique of analyte addition is uncommon in ICP-OES, the relevant software for its implementation is lacking in some software packages. It can nevertheless be easily applied using the following procedure. A quantity of the analyte, which corresponds approximately to the expected concentration in the sample, is added to the sample. In the case of doubt about the anticipated concentration, an addition that is too high should preferentially be chosen, because an addition that is too low can cause a higher error in the result. Of course, several additions are also feasible. The concentration of the quantity added, but referred to the total volume, is entered as the standard concentration. The spiked sample (the sample with the addition) will then be measured and treated within the context of the calibration procedure as if it were the standard. Next, the unspiked sample is aspirated and measured as if it were the blank solution. The calibration function of the software then calculates the slope for this matrix from these two values. Finally, the blank solution is measured as a sample. The calibration algorithm then calculates a negative concentration. The absolute value of this is the concentration in the unspiked sample.

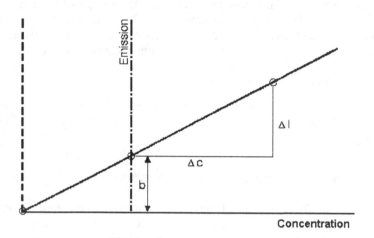

Fig. 115: In order to determine the concentration by the technique of analyte addition, the slope of the linear function ($\Delta I/\Delta c$) and the intercept b is determined. ΔI is the difference between the intensities measured for spiked and unspiked sample, Δc is the concentration added to the sample (amount added and calculated with respect to the total volume of the spiked sample), and the intercept b is the intensity of the unspiked sample.

When applying the technique of analyte addition, the linear range must not be departed from under any circumstances, since otherwise a result that is too high will be calculated. Contamination in the blank solution will also cause the calculated concentrations to be too high. A background correction is required, since typically the background will change, just as the analyte sensitivity is influenced by the matrix.

If several samples of the same matrix have to be analyzed, the sample, which has been analyzed using the described procedure, can be used as a calibration solution for the subsequent samples. If the concentrations are so low in the sample solution that they cannot be measured with good reproducibility, the spiked sample can also be used as a calibration solution.

4.3.1.4 Further Measures to compensate for Non-spectral Interference

The use of a surfactant to compensate for non-spectral interference is successful only in those cases where the non-spectral interference is produced by a difference in surface tension between calibration solutions and samples. In such a case, the addition of a surfactant is a kind of extended matrix matching.

The addition of an ionization buffer avoids excitation interference, which is found quite frequently (particularly with axial viewing). There is a risk of contamination, since larger quantities need to be used. In addition, the buffer must be added to all solutions in contrast to matrix matching where this operation is confined to the calibration solutions.

4.4 Optimization

Optimization is that part of method development aimed at obtaining the best possible results. A prerequisite for method optimization is the correct adjustment of the instrument (e.g. alignment of the optics with the analyte channel).

In ICP-OES, optimization is not always easy because

- In the typical case of a multi-element analysis, the optimization gets very complex
- Most parameters are mutually dependent (e.g. the optimal viewing height changes when the carrier gas flow is varied)
- It is often difficult for the user to decide which optimization goals to go for and what compromises are linked to an optimization goal (e.g. when optimizing for sensitivity, robustness may degrade).

An elaborate optimization makes sense only for "difficult" elements and samples. In many cases, good results can be reached using the default parameters provided by the instrument manufacturers. Only if the analytical performance is inadequate should one try to improve it by optimizing. During optimization, one should look not only at an isolated issue but consider the analysis as a whole. For example, there is little use in gaining, say, 10 % better sensitivity if the long-term stability deteriorates dramatically. Another example would be slightly improving the limit of detection while at the same time extending the time required for the analysis. Alternatively, the performance for one element improves, while for five others it becomes inferior. Of course, in this last example it may still be sensible to improve the results for the one element if that element is particularly important and sufficient "reserves" exist for the five others.

Even if it were in principle possible to measure every element or every analytical line at optimal conditions with your spectrometer, this ability should be used with utmost restraint. Changing the plasma conditions will often (automatically) upset the system, and any potential advantage may be ruined by this. Furthermore, a considerable loss of time will be caused by constantly adjusting to the new conditions. This delay is usually not acceptable in routine analysis. Finally, small differences have little influence on performance (e.g. RF powers of 1140 W and 1200 W give comparable plasma conditions). Therefore, the aim should be to bundle emission lines with similar excitation characteristics into groups and find group compromise conditions for these elements. As a final remark, it should be pointed out that excitation interference increases using a so-called "cold" plasma.

4.4.1 Optimization Goals

Optimization goals are very diverse. While maximum sensitivity is striven for in one laboratory, another laboratory may be concerned about increasing the analysis speed. Frequent optimization goals are
- Limits of detection
- Sensitivity (BEC)
- Precision
- Long-term stability
- Freedom from spectral interference
- Robustness against non-spectral interference, particularly excitation interference
- Analysis speed.

As well as the performance characteristics which are described in more detail in Sect. 4.5 "Validation", another feature is typically also of very great interest. This is a quantitative attribute: the analysis speed. This depends on a number of parameters that will be only listed here. These parameters depend on other characteristics of the spectrometer. They must be determined individually for each method for each particular instrument:

- How long does the sample take to reach the nebulizer?
    - Pump rate
    - Interior diameter of the peristaltic pump tube
    - Viscosity of the sample
    - Adsorption of the analytes on the tube material

- How long does it take for the aerosol to be replaced in the nebulizer chamber?
    - Properties of the nebulizer chamber surface
    - Velocity of the carrier gas flow
    - Volume and surface of the nebulizer chamber
    - Adsorption of the analytes at the nebulizer chamber

- How long does it take the plasma to reach a stable signal after the sample is aspirated?
  - Power regulation and reserves of the RF generator
  - Relationship between norm temperature and plasma temperature

- How long must the signal be integrated in order to obtain the desired reproducibility?
  - Expected reproducibility
  - Intensity of the emission
  - Noise components from sample introduction system and plasma
  - Noise components from optics and detector

- How many elements have to be determined?
  - Irrelevant for truly simultaneous spectrometers

- How long do I have to wait before the remains of the previous sample cease to affect the result for the current solution?
  - Tolerable carryover effects
  - Adsorption of the analytes in the sample introduction system
  - Dried material stuck to injector and torch.

The technological advances are also reflected in the speed of the analysis. For comparison purposes, the analysis time for 10 elements at different concentrations using 3 repeat measurements was recorded. To compare the analysis times, the default values suggested by the same manufacturer (PerkinElmer) were used. The rinse times were chosen so that good signal reproducibility and minimal carryover were obtained. In 1980, the analysis time was 13 min with the Czerny-Turner sequential instrument ICP/5000, in 1985 it was 7 min with the double monochromator system Plasma II, and, in 1997, 3 min with the array spectrometer Optima 3000.

4.4.2 Optimization Parameters

After the analytical line is selected and the processing parameters are fixed, the main parameters which the user can influence are the excitation conditions in the plasma. These depend on
- RF generator power
- Nebulizer gas flow
- Auxiliary gas flow
- Plasma gas flow
- Pump rate
- Viewing height for radial viewing.

4.4.3 Optimization algorithms

There are several strategies to search for the optimum:
1.   Trial and error
2.   Complete or partial determination of the parameter-effect relationship
3.   Use of optimization algorithms like SIMPLEX.

To list trial and error as an optimization algorithm may appear provocative. However, it would appear that this is the most frequently used algorithm. If an optimization is performed at all in the routine laboratory, typically only one parameter is changed. If an improvement is observed, then the conditions are usually left at this, as a thorough study of the complete parameter-effect relationship is often not practicable for most parameters in a routine laboratory. In order to make an optimization feasible in such a situation, elaborate strategies have been developed to enable the user to quickly get to the optimum. All strategies follow a similar procedure. At first, the region is covered in relatively big steps, which gives a rough idea of where the optimum might be. Then this is located more precisely by taking smaller steps. Care has to be taken in this procedure that the parameters are independent of each other. Since this cannot be ruled out (in fact it is quite probable), a thorough optimization is usually carried out iteratively in several steps.

✄   **The setting of the nebulizer gas flow and the radial viewing height depend on each other. How then can I optimize both?**

First check in which range the best height might be approximately located by passing over the analyte channel in 3 mm steps. The best viewing height is then set (if necessary by interpolation). The nebulizer gas flows are then changed in steps of 0.1 L/min. Take the two best settings and vary the flow rate between them in 0.01 L/min steps. The optimum nebulizer gas flow is then used in the search for the best viewing height in 1 mm steps in a range of ± 3 mm.

Another opportunity to arrive at optimal conditions with little effort consists of the use of elaborate optimization algorithms. Of these, the SIMPLEX algorithm is the best known [316]. (The outline of its functionality is described in the box below.) There are modifications of the SIMPLEX optimization routine [317]. A variation, which was implemented in a commercial instrument was known as the "directed search", which uses more starting points [318]. Although such algorithms will quickly find an optimum with considerable safety, they have not been able to gain widespread acceptance in the routine laboratory.

📖 **How does the SIMPLEX algorithm work?**

Here, only the principle of the procedure will be briefly described. Detailed instructions on exactly how to use the algorithm should be obtained from the literature [319].

The best way to understand the optimization algorithm is by analogy: Imagine a landscape with a hill in it. The hilltop is the optimum sought after. The optimization task consists of climbing to the hilltop. However, due to heavy fog, the visibility is zero. (Just to make you feel comfortable, rule out the possibility that there could be any steep slopes, fences, creeks or other obstacles.) You start anywhere in this landscape, and the only thing you can to do orientate yourself is to sense the slope. You could do this by going sideways in two directions, and feel where it goes up and where it goes down. Of these three points, one will be the lowest. It is then plausible that a higher point lies on the other side of the line connecting the two other points. This highest point and the two starting points form a triangle. This is now the new base for a renewed check of the slope. Again, there will be a new higher point, which in conjunction with the two next highest points, form a new triangle. This is used to project to an even higher point... You continue upwards until the difference in altitude is so small that you can assume that you have arrived at the top of the hill.

4.5 Validation

The validation concludes the method development. Its purpose is to
- Assure accuracy
- Prove the specificity
- Estimate the typical precision
- Determine the limits of detection
- Determine the limits of quantification
- Specify the working range
- Determine the robustness.

At this point, it should be made clear that instrumental measurement errors are only a part of the total error. All other steps in the analytical process are considered to be much more subject to error, as indicated in Fig. 116.

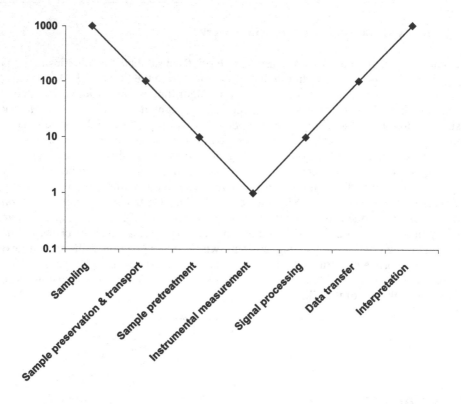

Fig. 116: Rough indication of the relative importance of typical analysis errors.

4.5.1 Accuracy and Specificity

To state that a method is accurate means that, through its use, no systematic errors and only small random errors are made [320]. In ICP-OES, accuracy is closely associated with specificity, since the determination of an element is carried out using its emission wavelength. Spectral interference causes a non-specific signal on the analytical line, causing erroneous results. Accuracy also depends on the absence or successful correction of sources of non-spectral interference. The absence of any interference must be proved and documented.

Measures to ensure accuracy vary according to the circumstances and possibilities. In principle, there are three possibilities, which may be combined for optimal security:

1.   Use of well-characterized materials
2.   Comparative measurements using independent techniques
3.   Systematic exclusion of errors.

1. The analysis of pre-analyzed samples, such as standard reference materials or samples from a round robin, can be utilized for validation. Occasionally, laboratory standards are used. This could be a step to assure accuracy, but the opposite could be equally true. It depends on how much effort was used to arrive at the concentrations for this laboratory standard. If this was subject to a critical examination then it may very well be acceptable. If possible, however, standard reference materials should be used, because their concentrations were established by a number of expert laboratories with different analytical techniques [321, 322]. The number of certified elements in such materials is limited. Therefore the concentrations of further elements are sometimes published independently [323].

2. Since the choice of sufficiently quality-assured, pre-analyzed samples or elements is frequently limited in standard reference materials, a comparison with results that have been determined by an independent technique can provide important evidence that no systematic errors are being made with the method.

3. The systematic exclusion of errors requires verification that there are no errors falling under any of the following headings:
   - Spectral interference
   - Non-spectral interference
   - Calibration.

It is very important that all three of the last group of possible error sources should be meticulously scrutinized. In order to exclude spectral interference, other undisturbed wavelengths of the analytes should be considered (if these are available). If significant differences in the results are noticed, the spectra should be checked very thoroughly. The absence of non-spectral interference should be checked by spiking, by diluting (provided that the sensitivity makes this feasible), and by then showing that the recovery rate is good. Finally, calibration solutions produced independently should be used (if possible from different manufacturers).

The validation is generally the last stage of method development. Nevertheless, in some cases it can make sense to constantly validate the method by checking results during routine analysis. This is especially helpful if the matrix composition of the samples varies a lot (e.g. some types of environmental samples), and the impact of the results is crucial. In such a case, one or more samples should be spiked or diluted (or both). A good recovery rate will signal that the results can be trusted. In addition, the safety of the analysis can be increased by the use of several analytical lines for one element.

4.5.2 Reproducibility

The reproducibility is a measure of the short-term stability. It is equivalent to the relative standard deviation (RSD) or the variation coefficient. The short-term stability is influenced by two factors [324]:

- Fluctuations in the sample introduction system and in the plasma ("flicker noise")
- Noise of the detector ("shot noise").

The reproducibility r should not be confused with the comparability R. The reproducibility refers to the fluctuations of the instrumental measurements (with the same operator), while the comparability reflects the differences between different instruments (and operators).

Even though the reproducibility should not be taken as a measure of the accuracy, it is nevertheless an important indication of the reliability of the result. An insufficient reproducibility, particularly when the number of repeats is very small, will make a correct result a "stroke of luck" [325]. Since one should not rely on luck (especially in analytical chemistry), one should make every effort to determine the result with good reproducibility.

The use of reproducibility as a performance criterion for the instrument only makes sense if other contributing factors that do not represent the state of the instrument are ruled out. The most important factor is the concentration of the analytes. Here it is important to use concentrations where one can anticipate a reproducible signal for statistical reasons. A concentration of more than 1000 times the detection limit will give an intensity which will give information about the character of the instrument. Table 8 records the improvement in the reproducibility of commercially available ICP-OES instruments over the time of their use in routine laboratories.

Table 8: Typical ranges of reproducibility from three repeats (using PerkinElmer ICP emission spectrometers)

Year	Instrument	RSD [%]
1980	ICP/5000	1 ... 5
1985	Plasma II	0.5 ... 2
1993	Optima 3000	0.2 ... 0.7
	Using simultaneous internal standardization	0.02 ... 0.2

⚒ **I expect better reproducibility. What can I do?**

Assuming that the "hardware" is OK (for hints on what to do if it is not, refer to Chapt. 6: "Trouble-shooting"), the reproducibility can be improved by increasing measurement times. If this is not successful, find out whether a steady-state signal has yet been reached. A longer rinse time before beginning the measurement also helps. If poor reproducibility is observed after a previous measurement of a solution with a very high concentration, then a longer rinse time between the samples may be beneficial.

The reproducibility will deteriorate from a statistical point of view (in accordance with law of error propagation) if a number of unrelated measurements are made for its calculation. For example, this is the case for the sequential background correction measurement in classical sequential and simultaneous spectrometers. Here, independent measurements are carried out for the peak and for the background correction points. Thus, a background correction causes inferior reproducibility, which is in this case accepted in order to obtain a better accuracy.

Since the fluctuations of the signals of peak and background result from the instability of the instrument (primarily from the sample introduction system), a truly simultaneous measurement, which is made possible by an array detector, can compensate for these fluctuations [326]. In this case, background correction will result in better reproducibility (and also accuracy). This is even more pronounced for very small signals where the intensity of the measured peak is close to the limit of detection. The improvement in the precision can be further increased by the simultaneous measurement of an internal standard (assuming that the norm temperatures of the two transitions match approximately). In such cases, a reproducibility below 0.1 % can be achieved (compare Sect. 7.3.4.3 "Noble metals").

4.5.3 Limit of Detection

A high proportion of the analyses performed by ICP-OES are in the trace analysis range. For this purpose, good detection capability is demanded. Here, not only is a good signal/background ratio important but even more so a good signal/noise ratio. In order to quantify the latter, the concept "limit of detection" is used. In contrast to the background equivalent concentration (BEC) , which is defined unambiguously, there are different views on the interpretation and determination of the "limit of detection" [327]. To start with, a definition that is generally accepted: the limit of detection is a statistical quantity expressing a probability that an analytic signal could be clearly distinguished from the background [328]. This does not imply that the signal can really be measured and thus be used for quantification. In other words, the limit of detection does not relate to the lower working range of a method.

There are two generally accepted procedures (and innumerable varieties of these) for determining the limit of detection. One is quite elaborate (and also encompasses other aspects), which is known as the method of calibration. The other is much simpler, because it essentially measures the fluctuations of the background when a blank is aspirated: this is termed the blank method.

The method based on a calibration line uses the area of confidence of a calibration [329]. In accordance with the specified operating procedures, 10 standard solutions are used whose concentrations increase in equal steps. It is a prerequisite that throughout this concentration range the homogeneity of variance has been checked. However, this prerequisite is not fulfilled over the complete working range typically used in ICP-OES. Therefore, the method would be strictly applicable only in a very restricted working range. Moreover, it is very time-consuming. In routine analysis, typically not many calibration standards are used, so that the method of calibration does not actually reflect the calibration used under normal circumstances. In addition, in order to get meaningful results, care must be taken to ensure that the concentrations of the solutions lie in a range close to the expected limit of detection. The utmost care must be taken in the preparation of the solutions, because one incorrectly prepared solution will ruin the whole series, and will thus lead to totally unrealistic figures for the limits of detection.

The blank method is more workable. Here the limit of detection is determined from the reproducibility of the measurement of the background (with at least 10 repeats). This method is preferred in practice since it is simpler to execute and the potential for errors is less. The limit of detection c_N using the blank method is calculated as

$$c_{LOD} = k \cdot s_{BG}$$

where k is the statistical factor for the probability and s_{BG} is the reproducibility of the background measurement in units of concentration.

Depending on certainty of the statement that the sample signal is detected, different statistical factors are used. After a long period of very controversial debate, nowadays the statistical factor 3 is frequently quoted, and this corresponds to a probability of 99.7 %. The factor 2 corresponds to a probability of 95 %.

ⓘ   **What can be quantified at the limit of detection?**

Nothing!

Often, even mentioning the term "limit of detection" can start very heated discussions – and more often confusion. Depending on the group discussing the topic, the term can have different connotations. A common misapprehension, especially with persons who normally have no close contact with (chemical) analysis, is that this concentration was actually measured. Then it can happen that the limit of detection is used for the calculation of a quantity of material. For example a detection limit value can be multiplied by x millions of cubic meters to arrive at amounts in the order of kilograms of a poisonous substance, where the laboratory has found "nothing".

Even though one cannot find anything, one can nevertheless have wonderful discussions about how to define this "nothing". Of the common techniques to determine the limit of detection, essentially there are the method of calibration and the blank method, with a number of variations. The method of calibration should certainly be regarded with the greatest caution. The great expenditure of effort necessary to arrive at the detection limit by the method of calibration might suggest that the result is reliable. However, the opposite is sometimes true, at least in the view of the author, who knows of examples where, rather than determining the limit of detection, the pipetting accuracy of the laboratory was revealed. (Sometimes the method of calibration produced numbers that do not make sense, e.g. limits of detection for ICP-OES and ICP-MS turn out to be the same!) Theoretically, the two methods determination of the limit of detection should lead to the same results. In practice however, this is not always the case. The values for the detection limit according to the method of calibration are often higher.

The blank method and its variations are the most commonly used ways to determine the detection limit because they are more practicable. For the blank method, there are a large number of ways of determining the numerical value of the limit of detection, depending on whether or not the concentration measured for the blank reading should be included, and which factor to express the statistical safety should be used. These are the main areas where differences become obvious. Very often the difference is not made clear, because it is not mentioned how the detection limit was determined and which statistical factor was used for calculation. Therefore, the value for the detection limit can increase twofold simply by the choice of the factor for the statistical safety. For quite a long time, the factor 2 was very commonly used for k. Some instrument manufacturers in particular preferred this, because its use reduces the limit of detection by a third compared to a value with the factor 3. Aside from the choice of the factors 2 and 3, the products of their multiplication by the square root of two are used by some. This factor arises for the case that at the determination of the limit of detection no background correction was applied. A factor $\sqrt{2}$ is then used to convert the limit of detection to a realistic measurement, where sequential background correction (the typical way of background correction for "classical" sequential and simultaneous instruments) is performed. Unfortunately, sometimes the factor $\sqrt{2}$ is also applied when background correction was used during the determination of the detection limit. Furthermore, this multiplier does not make any sense at all for a true simultaneous background correction (as with array detectors). Here, the noise of the background is reduced because of averaging out the coherent fluctuations of the instrument, thus lowering the limit of detection.

In order to avoid the confusion associated with the concept of "limit of detection", it seems reasonable, to scrap this term and replace it by the well-characterized terms "background equivalent concentration" and "base line noise" [330]. These terms do not suggest to anybody that a measurement can be performed at this level and it will eliminate the misuse of the statistical factor. Even if this suggestion should ever be adopted, some time will elapse before this happens. The author will therefore use the concept "limit of detection" for the time being because of the better comparability.

Since the introduction of ICP-OES, limits of detection have improved steadily, reaching lower concentrations (Table 9). Limits of detection depend substantially on the resolution [331], the light throughput of the optics, the sensitivity of the detector and its dark current (which should be low) [332]. Axial viewing of the plasma was the most important single step forward. The use of an ultrasonic nebulizer is another means of lowering the limits of detection by about a factor of 10 for samples with minimal matrix load. It should be critically observed here that when working in this concentration range an increasing amount of contamination is to be anticipated, so that care should be taken

Table 9: Decrease in limits of detection in ICP-OES over the course of the time using the example of the PerkinElmer ICP emission spectrometers ICP/5000 (1980), Optima 3000 (1993) and Optima 3000 XL (1997). All detection limits were determined by the blank method using the statistical factor $k = 3$ [concentrations in µg/L]

	1980 radial	1993 radial	1997 axial
As 193	150	50	5
Cd 214	3	2	0.3
Cr 267	5	2	0.2
Ni 231	10	5	0.7
Pb 220	50	10	0.8
Zn 213	2	1	0.1

Essentially, when determining the limit of detection in ICP-OES, the fluctuation of a small signal in the range of intensity of the background signal is measured. Consequently, the limit of detection is largely influenced of the height of the background. If the background is low, then the limit of detection is mainly determined by the sensitivity of the analytical line. At low wavelengths, the argon plasma emits a continuum with low intensity. In this wavelength range, the light throughput of the optics is crucial for good limits of detection. At a high background, the fluctuation of the background dominates the limit of detection. If the background is measured simultaneously with the analyte signal, then the correlated fluctuations of these cancel themselves out.

4.5.4 Working Range

An important reason for the commercial introduction of ICP-OES was its immense linear working range, which can extend to up to six orders of magnitude. This range is rarely reached in routine analysis, since the carryover of elements present in very high concentrations on consecutive sample with low concentrations typically reduces the working range to about four orders of magnitude.

The working range starts with the limit of quantification c_{LOQ} at the lower end. The limit of quantification can be determined along with the limit of detection in the same procedure. When determining the limit of detection by the blank method, three times the limit of detection (at $k = 3$) is usually regarded as the limit of quantification. It is often calculated in simplified way as

$$c_{LOQ} = 10 \cdot s_{BG}$$

The upper end of the working range is set by the calibration solution with the highest concentration. Otherwise, it stops at the end of the linear range at the latest (except in the rare case that a non-linear calibration function is used).

In order to check the linearity, at least 10 solutions are prepared that cover the anticipated range. Their corresponding intensities are measured. If only a small range needs to be checked, the concentrations prepared should be in equidistant steps (e. g. 2, 4, 6, 8, 10, 12,...). In order to check the linear range for a larger working range of several orders of magnitude, which is typical in ICP-OES, the steps should be "equidistant on a logarithmic scale" (e.g. 1, 2, 5, 10, 20,...). Of the many alternative ways to evaluate the linear range, three procedures are described here.

According to EPA guideline 200.7, one calibrates with ascending concentrations. The next higher concentration is measured as a sample, and if the calculated concentration is not less than 10 % of the anticipated concentration for this solution, the solution is then used as a calibration solution for an extended calibration. This procedure is continued as long as a solution gives less than 10 % of the prepared concentration. The end of the linear range is set at 90 % of this concentration [333].

As a graphical variation, the concentrations and intensities are converted to their logarithm values and entered in a diagram with double logarithmic scale. Caution is required, since in this manner of representation slight curves appear to be straight lines. A very critical examination is necessary in this case!

For a statistical assessment of the linearity, regressions of the first order (linear function) and of the second order (quadratic function) are calculated. With the help of statistical tests, the function that best describes the measured concentration-intensity correlation is found. For details, refer to the literature [334].

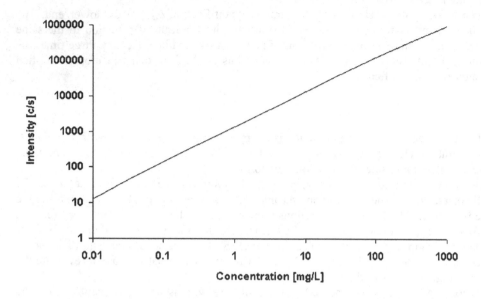

Fig. 117: Check of the linearity of Sn at 189 nm with axial viewing. The graphic evaluation of the linearity test is based on visual judgment of the points in a double logarithmic representation. The form of display "distorts" in such a way that one gets the impression that the function is linear although it might be slightly curved. On very close inspection it can be seen that the linearity ends at about 100 mg/L. .

✂   **I have checked the linearity of 20 elements and on no account can I work to 1 g/L even though I use lines of low sensitivity!**

The check of linearity should be performed with single-element solutions. If one uses a multi-element solution (e.g. there is a commercially available calibration solution which contains 23 elements at a concentration of 1 g/L) and dilutes this, the linearity appears to be insufficient. However, the effect observed is not related to the linearity issue, but is caused by the matrix effect (non-spectral interference) in a solution containing components at such high concentrations.

4.5.5 Robustness

In a proof of the robustness, it should be shown that small variations in the analytical procedure used in the instrumental measurement will not have any significant influence on the result. This is particularly difficult to prove, since a large number of parameters can influence a result. Often one is not even conscious of the contributing effects. Examples of parameters that can easily be underestimated are the barometric pressure, which can cause wavelength drifts in certain systems, or the influence of the atmospheric humidity, which in certain situations can cause water to condense on the induction coil, which can affect the RF power coupled to the plasma. The test for robustness is therefore hardly feasible in practice. Information on deviations is however provided by quality control tests (compare Sect. 5.3 "Quality assurance"). Quality control could be considered to be an ongoing test of the robustness of the method. In the practice, testing for robustness is hardly ever done.

The "robustness of the method" should not be confused with the "robustness of the plasma" (compare Sect. 2.2.3 "Spectroscopic properties of the ICP"). A robust plasma is a necessary prerequisite but not a guarantee of a robust method.

✂   **ICP-OES: Method development – step by step – a short outline**

1. Make sure that the spectrometer is in the optimal operating state

2. Select wavelengths for the elements (preferably several per element)

3. Generate spectra of at least the followings solutions:
   a. Calibration solutions
      For example, multi-element solution(s),
      Blank solution (measure several times if necessary), and
      Single-element solutions of the analytes and potential interferents if appropriate
   b. Representative samples,
      Possibly also samples differing from the collective for a "more universal" method
   c. Quality control samples

4. Show spectra in the graphic display mode and evaluate for usability

5. If necessary, change the scale in order to obtain a better evaluation of the background

6. Set background correction points (preferably two on each side)

7. Measure again or, if possible, recalculate data and
   compare the results for one element at several wavelengths

8. Choose one suitable wavelength per element
   In some cases, it may be beneficial to use several wavelengths in the routine

9. If there is no undisturbed analytical line, correct if possible by
   a. Multivariate regression techniques
   b. Inter-element correction

10. Check whether there is non-spectral interference, and if necessary correct by
    a. Matrix matching
    b. Internal standard

11. Optimize the excitation conditions if necessary

12. Validate the method
    Determine the analytic performance criteria, and if necessary repeat step 11.

5 Routine Analysis

Routine analysis constitutes by far the highest proportion of everyday laboratory work. Before samples are measured using an ICP-OES instrument, a number of operations must be performed [335]. All these should be documented in Standard Operating Procedures (SOP). A simplified form listing standard operating procedures for starting up an ICP instrument is shown in Fig. 118 .

| AnalytikSupport | Laboratory XYZ ICP-OES # 123 | Page: | 1 |
| | | Date: | 2002-07-24 |

(1)   Turn on exhaust system

(2)   Turn on recirculating cooler

(3)   Check Argon supply (minimum pressure 6 bar)

(4)   Open Argon valves

(5)   Turn on PC, screen and printer

(6)   Start ICP software

(7)   Check tubes of peristaltic pumps for wear
        and change if necessary

(8)   Insert tubes into peristaltic pump

(9)   Renew rinse solution

(10) Put sample capillary into rinse solution

(11) Ignite the plasma. Watch the ignition procedure

    Attention:
    If there is a loud noise associated with the ignition,
    turn off immediately!

(12) Wait for 30 min warm-up time of the plasma

Fig. 118: Example of standard operating procedures (SOP).

All operating steps of the analytical procedure will influence the result. While most of these are normally carried out in the analytical laboratory or in close coordination with it, the interpretation of the results is usually done elsewhere. Sometimes, the data are not evaluated or interpreted meaningfully. Especially if this is the case, a considerable error can be made [336]. The most important operating steps are:

- Sampling, in order to get a representative sample of the whole. Inhomogeneity must be taken into consideration [337, 338]
- Sample storage and stabilization of the analytes
- Transport to the laboratory
- Taking of a laboratory sample from the original sample
- Specific sample pretreatment in order to convert it or the analytes into a form that can be used by the analytical instrument
- Measurement of the physical information value
- Conversion of the signal into a concentration (processing)
- Data transmission, preferably electronically (to minimize transfer errors and workload)
- Interpretation of the data.

5.1 Preparation

5.1.1 Sample Preparation

The sample preparation steps are usually performed by the same persons who operate the instrument. Sample preparation is usually closely tied to the specific application (further details may be found in Chapt. 7 "Applications"). Only some generally valid points will be mentioned briefly here.

In general, the lower the concentration of the matrix, the lower is the risk of interference. With respect to digestion reagents, practical experience has shown that the fewest problems occur when using nitric and/or hydrochloric acid. Of course, the exception proves the rule here also! Digestion reagents that lead to increased viscosity, such as sulfuric and phosphoric acids, are not recommended, since their use leads to strong signal depressions. Besides the potential health hazard, hydrofluoric acid also has analytical disadvantages. It attacks glass and quartz, of which at least some parts of the sample introduction system are made in many instruments. This attack can be prevented by masking with saturated boric acid solution. However, there are a number of volatile and insoluble fluoride compounds.

5.1.2 Warm-up Time

On ignition of the plasma, it heats the torch, the sample introduction system and the torch compartment. The sensitivities of the respective analytes change during this period. Therefore, time must be allowed for the warm-up before starting the analytical measurements. However, during this time other jobs can be done such as checking spectra. The warm-up time depends on the emission line, and important factors in this context include the relationship between the temperature characteristics of the emission line and the temperature of the plasma, the torch and (depending on the instrument) the sample introduction system. The temperature of the latter two changes because of the thermal radiation from the plasma. Since the resulting signal change proceeds asymptotically to a final value, the warm-up time depends also on the acceptable deviation from the final signal.

5.1.3 Delay and rinse times

A quantitative measurement in ICP emission presupposes that a stable signal has been reached ("steady state"). This takes quite some time and consists of several consecutive steps. First, the sampling tube and pumps must be filled with the sample. The time necessary for this depends on the pumping rate and the length and diameter of the tubes. Next, the sample is nebulized. The aerosol formed spreads out in the nebulizer chamber. The gas flow rate, the nebulizer volume and the material of the nebulizer chamber determine the duration of this step. The sample aerosol finally reaches the plasma. One must now wait until a new temperature equilibrium is established. This depends on the matrix (composition and concentration) and the plasma conditions.

During this time, the instrument waits before taking measurements. The period between dipping the sample tube into the sample solution and the beginning of the measurement is the **delay time** (or "read delay"). Following the measurement of the sample, the remainder of it must be rinsed out of the sample introduction system except for an acceptable amount left behind. This is the **rinse time** or wash-out time. (Note that the rinse solution used should not be the blank solution!!!) Frequently, 1 % or 0.1 % of the steady-state signal is used to indicate the threshold value for the acceptable remainder. Of course, the choice of the threshold value determines not only the rinse time but also the working range.

The delay time should be determined experimentally for a given application. A practical procedure is to choose a very large number of repeat measurements (e.g. 100), set the delay time to zero for this experiment and dip the tube into the sample while simultaneously starting the measurement. The total duration of this experiment (for the 100 repeats) is registered. The measured intensities are plotted in a diagram in which the intensity is recorded as a function of the repeat number (or the time), as illustrated in Fig. 119. The diagram shows the time at which the signal ceases to increase. This is the delay time.

The delay time can be accelerated by filling the tube leading to the nebulizer quickly by using a fast speed of the peristaltic pump. This can be triggered automatically in some spectrometers. This high pump speed should be reduced again as soon as the solution reaches the nebulizer, as it is used only for the fast filling and would simply waste the sample material if it continued. After this stage, other factors determine the duration of the delay time. If rapid rinsing is used until immediately before the measurement is started, the analytical signal can be destabilized.

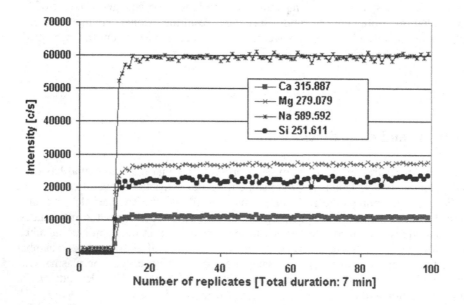

Fig. 119: Experimental determination of the delay time. The example shown here is the delay time for the slurry technique. (Generated with a PerkinElmer Optima 3000)

It takes a while for the remainder of the old sample to be washed out. The duration of the rinse time also depends to a great extent on the element. Some elements have unusually long residence times in the sample introduction system (e.g. the elements B and Zn). This is also known as "memory-effect" . The rinse-out time is determined experimentally, analogously to the determination of the delay time, by a sequence of measurements over a period (Fig. 120). The time taken for the accepted concentration level of the remaining analyte to be reached is taken as the time for rinsing between samples.

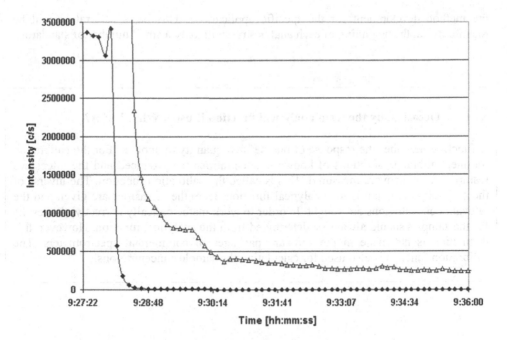

Fig. 120: The rinse-out behavior shown for the example of Zn. This element is rinsed out rather slowly from the sample introduction system. A few seconds after the sample tube is removed from the sample and placed in the rinse solution, the signal is so small that it is no longer visible in the normal scale of this diagram (black diamond-shaped symbols). Therefore, the intensity was multiplied by the factor 100, and these data are shown as triangles. One can see from this that the reduction of the original signal to 0.1 % of its initial value takes about 4 min.

5.2 Calibration

ICP-OES is a relative analytical technique. This means that the correlation between intensity and concentration must always be determined before the beginning of the analysis [339]. This correlation is described as a calibration function.

Calibration is necessary since the intensity at a given concentration depends on a number of parameters, not all of which can be determined. In applications where the exact result is not required (e.g. screening analysis or for a decision regarding the classification of a material), one can work with stored calibration data. For some simultaneous spectrometers, mainly those which are designed for use in the steel industry, a far-reaching basic calibration involving a multitude of standards is carried out by the instrument manufacturer, usually on installation of the instrument (or as part of the method development for the specific application). This basic calibration will be recalibrated at the beginning of each analysis run with only a small number of standards.

📖   **Occasionally the term analytical function is used. What is this?**

A function describes the response of one defined quantity to another. For the calibration routine, calibrating solutions of known concentrations are produced and the intensities resulting from them are measured. This is called the calibration function. The inverse of the calibration function is the analytical function. Here the intensities are given and the relevant concentrations are sought. In order to work mathematically correctly, the results for the samples should always be determined from the analytical function. However, this distinction is not made in the software packages of commercial spectrometers. The calibration function is also used for calculating the sample concentrations.

5.2.1   Calibration Solutions

5.2.1.1   Number of Calibration Solutions

The relationship between concentration and intensity is linear up to six orders of magnitude. Therefore, in many cases two calibration solutions are sufficient: an upper calibration solution and a lower calibration solution (usually the blank solution). The results achieved with two standard solutions are no different from those achieved using several standards provided that there are no other (matrix) influences. However, if one opts to work with two calibration standards, it is mandatory that the obtained calibration function should be checked against samples prepared independently (e.g. quality control samples).

Frequently, a series of several calibration solutions is prepared by dilution of a stock solution. The calibration functions so obtained do not guarantee a true calibration since the stock solution could also have been prepared wrongly or could have changed. The use of several calibration solutions encourages the feeling that one has done a good job in ensuring the accuracy (and the correlation coefficients look great!) and that one has no need to test the validity of the calibration using a sample prepared independently. Consequently, this procedure may be less effective in ensuring accuracy.

ⓘ   **How many calibration solutions should I use?**

Not only are limits of detection continually debated, so also is the question of the number of standards to be used in ICP-OES – passionately and always without any conclusion. There are advocates for two, three, four, five or even ten calibration solutions. Moreover, everyone is right in a way. If all the solutions are prepared correctly and there is no interference, the number of solutions for calibration is irrelevant. Software packages which store all spectral information and which allow the user to reprocess the data in a variety of different ways, show that for identical data sets the results are comparable whether one uses a blank solution and a calibration solution or a series of ten calibration solutions.

If one uses only two calibration solutions, one takes a certain risk, which similarly exists, however, if one uses a series of ten solutions: The calibration solutions can be prepared wrongly. However, this can happen in just the same way to the series of ten solutions, because these solutions are frequently prepared as a series by dilution of a stock solution. In such a case, one has a false sense of security. To avoid the danger of an erroneous calibration function, the only way to be sure is by checking it with a (QC control sample) solution prepared independently. This is necessary, whether one calibrates with two, three or ten solutions.

The bottom line is: put your effort into carefully preparing the calibration standards rather than measuring a series of calibration solutions that are prepared with too little care.

5.2.1.2 Concentrations in Calibration Solutions

The concentration in the upper (most concentrated) calibration solution should be chosen so that it can be measured with good reproducibility. This means that the concentration should at least be around the factor 100 above the limit of detection.

In addition, the concentration in the upper calibration solution should be at the upper limit of the expected concentration range of the samples, since an extrapolation is always more prone to error, especially over a big range. Figure 121 illustrates this fact. An error during the measurement of the blank solution causes a clearly incorrect calculated numerical value for the sample concentration if the concentration of the calibration solution is lower than the concentration of the sample. The same error for the blank solution causes a fundamentally smaller error for the sample result if the concentration of the calibration solution is greater than the concentration of the sample. Needless to say, all error sources should be avoided.

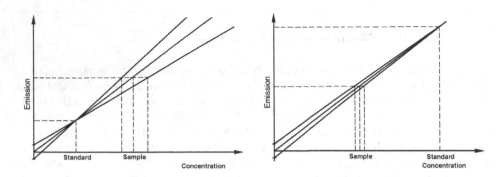

Fig. 121: The concentration of the most concentrated calibration solution should be greater than the concentrations expected in the samples. If the concentration of the calibration solution is lower than that of the sample, a small error in the blank reading causes a large error in the calculated concentration of the sample. When calibrating with a standard with a concentration greater than the anticipated sample concentration, the error is much smaller. In this example, only the error on the blank measurement is considered. However, variations in the calibration solution must also always be reckoned with. For that case, the argument and the necessary graphics become more complicated, but the conclusions remain the same.

Practical experience shows that good results can also be achieved with samples only a little more concentrated than the upper calibration solution. In the end, this limit is determined by the reproducibility of the measurement of the upper calibration solution. Care has to be taken in the case of the usual linear calibration function that the upper standard does not exceed the linear range. A further consideration is that the concentration in the upper calibration solution should be so low that no carryovers are observed. The latter is particularly important for elements with a distinct memory-effect (such as the element B).

As a rule, the lower end of the working range is not identical with the concentration in the lower calibration solution because in general this is the blank solution. In this case, the lower end is set at the limit of quantification (as a rule: $10.s_{BG}$).

For other analytical techniques, it is sometimes recommended to always set the concentrations in the upper calibration solution at the upper end of the anticipated concentrations in the sample. This guideline makes sense if the expected concentrations in the samples are so high that they can be measured with good precision. However, if the concentrations in the samples are very low, then this will result in the calibration being established with solutions which are measured with poor precision. The calibration function will likewise have a high degree of uncertainty. Since this shortcoming, however, is not taken account of in any instruments software known to the author, the reproducibility of the sample as determined by the system seems unrealistically favorable. Only if the statistical error is effectively propagated when calibrating will this show up in the reproducibility of the results and be obvious.

In ICP-OES, with its large linear range, one should prepare the standard with concentrations that can be measured with good reproducibility. Because of the large linear range in ICP-OES, higher concentrations in the standards will ensure that the correct slope will be found. Finally, calibration with a more highly concentrated calibration solution and interpolating to lower concentrations in the samples yields good results.

5.2.1.3 Multi-element Calibration Solutions

Preparing a multi-element calibration solution requires great care if more than 10 elements have to be pipetted into one calibration solution. This has been found by a many workers apart from the author. It is less demanding and also preferable, according to the concept of traceability, to use commercially available solutions. If you want to prepare multi-element calibration solutions, you should think beforehand of all the steps to be taken, gather together all the necessary items to be used (e.g. pipettes and tips, volumetric flasks, labels and pens, containers for the prepared solutions...), and set down in writing all the dilution steps, exactly how they are to be done and if necessary which intermediate dilution steps have to be taken. To summarize: all the steps of the operation must be thought of and planned exactly before you start mixing a multi-element solution. It is furthermore recommended that this work be done only when one cannot be disturbed by others.

The stock solutions used for preparing the calibration solutions should be as pure as possible. In general, the standards for ICP should be of higher purity than those for AAS (or, in order to be on the safe side, where any information about the use of the standard solution is lacking). In the case of the better quality stock solutions and multi-element calibration solutions, a certificate concerning the contaminants will accompany each solution. High purity is particularly important if different concentrations are mixed together in the calibration solution. The contamination risk is particularly high in the case of element solutions that are added in higher concentrations.

When using commercially available calibration solutions, the manufacturer will guarantee the content with respect to the analytes. Typically, the solution is accompanied by a certificate giving the measured concentrations of the analytes.

5.2.1.4 Multi-bottle Calibration

It is known that silver will precipitate in a hydrochloric acid solution. Frequently the acid anion and the acid concentration decide the stability of the calibration solution. If different elements are mixed in higher concentrations into an already existing calibration solution, insoluble compounds such as $AgCl$ or $BaSO_4$ can precipitate. Furthermore, there are a large number of compounds of low solubility which precipitate spontaneously once the corresponding ions are added to a calibration solution. Since not all the elements can be kept in solution alongside each other, most software packages have the ability to put together a calibration using different calibration solutions (multi-bottle calibration). Elements that lead to the formation of unacceptably low solubility compounds are pipetted into different standard solutions and the ICP-OES system will calibrate from these separate solutions, finally creating calibration functions for all elements.

5.2.1.5 Stability of Calibration Solutions

The calibration solutions have a finite shelf life. A generally valid statement is: the lower the concentration, the lower the stability. However, the element under consideration and the container material will influence the stability.

Solutions, which contain concentrations in the mg/L range are usually stable for months. Studies show that most elements maintain their original concentration after one or more months even at a concentration of 100 µg/L or below [340, 341]. Exceptions are Ag, Au, Pb, and Ta (removal by adsorption at the container walls and, in addition for Ag, photo-oxidation) and B (in the case of very low concentrations, increase by leaching from the glass containers used).

Generally, synthetic materials adsorb less than glass, which tends to prevent decrease of the concentration over time or carryover on multiple usage. The best materials are high-density polyethylene (HD-PE), polypropylene (PP), polytetrafluorethylene (PTFE) and perfluoroalkoxy (PFA) [342]. As always, the opposite is sometimes true; some noble metals tend to absorb to some types of synthetic materials.

5.2.2 Calibration Functions

As a rule, the calibration function is a linear (straight line) function [343, 344]. If several calibration solutions are used, most spectrometer software packages allow the choice among a number of calibration functions. Most functions rely on a regression. As a rule; the slope, calculated intercept and correlation coefficient are given as performance criteria [345]. Even if these performance criteria suggest that the calibration is of good quality, a number of error possibilities are nevertheless present [346], such as an inaccurately prepared calibration solution. The calibration therefore should be checked in every case with a solution prepared independently.

Most emission lines show a linear concentration-intensity relationship over a range of up to about six orders of magnitude. In the range above this, the function starts to level off because of self-absorption. Only a few elements (mainly the atomic emission lines of the alkali and alkaline earth metals) already show non-linearity at quite low concentrations. In this case, a non-linear function (2nd order) can enlarge the working range meaningfully. However, one should then check very precisely whether the calibration function of the spectrometer software correctly reflects the actual situation. A large linear range cannot be extended by the application of a non-linear calibration function, as is clear from Fig. 122. Instead, a combined function must be used in this case, with a linear function for the lower range and a non-linear function for the upper. Many software packages do not allow this. Then either the working range must be reduced or else an additional calibration of the measured intensities should be carried out with a separate program as a post-analysis data treatment [347].

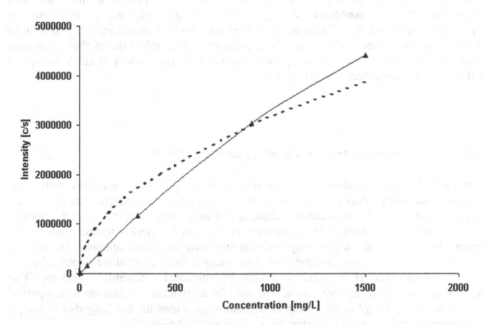

Fig. 122: A measured non-linear concentration-intensity relationship   (indicated by triangles) is linear in the lower concentration range, in this example to about 300 mg/L. The slope then decreases continuously above this concentration. A (calculated) quadratic function (dashed line) is already curved at the start of the lower range. The two functions, the calibration function determined experimentally and a quadratic function, are not congruent, which is imperative in order to use it for calibration.

5.2.2.1 External Calibration

By far most frequently used calibration technique in ICP-OES is external calibration. The calibration solutions are measured at the beginning and the calibration function is established. The samples are measured, and this calibration function is then used for the calculation of the concentrations in the samples.

There are differences in the exact way the calibration functions work. The main difference is in treatment of the function for the blank (at zero concentration). There are variations according to whether the blank intensity reading is subtracted from all consecutive intensities or the regression is forced through the blank reading or through zero intensity. Forcing through zero makes sense only if the intensity for the blank solution is also subtracted (which is superfluous if background correction is carried out, which as a rule is necessary). Forcing through the zero can cause great errors, especially in the upper concentration range.

In some software packages, different weightings are applied to the standards. Sometimes the lower concentration range is more strongly weighted, in other cases the upper concentration range. If weightings are applied, the user should inform himself how the calculation is done in this case. The experience of the author shows that the normal regression is optimal for most applications (and if one works with a two-point calibration, the issue becomes irrelevant).

5.2.2.2 Calibration by Analyte Addition (Standard Addition)

The method of analyte addition (also referred to as the method of standard addition) is used occasionally if other possibilities for the correction of non-spectral interference are not effective. Since this calibration technique will only yield correct results if a number of prerequisites are fulfilled [348], and since in the case of a multi-element analysis it is quite laborious, its use is generally not recommended in routine analysis. However, a meaningful use can consist in determining one sample from a uniform sample collective where the non-spectral interference cannot be corrected efficiently otherwise. This sample (or the spiked sample) then can be used for an external calibration with a perfect matrix matching for all further samples. Further suggestions for handling this technique are described in Sect. 4.3.1.3 "Calibration with Analyte Addition".

5.2.2.3 Bracketing Calibration

The technique of bracketing calibration is used very rarely for calibration in ICP-OES. In bracketing, two calibration solutions, a lower and an upper calibration solution, with concentrations grouped closely around the anticipated sample concentration are used. Originally, the bracketing calibration technique was utilized to match non-linear ranges of the calibration function with a linear function portioned into small segments. Figure 123 shows that this scheme obviously works best if the concentrations of sample and calibration solutions differ only slightly.

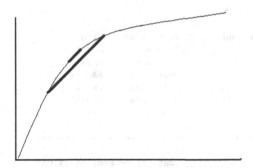

Fig. 123: The concentrations of the calibration solutions are fixed so they are very close to each other and very close to the sample concentration. In a curved calibration function, a linear interpolation with calibration standards close together yields a better match and a satisfying result. For greater distances between the concentrations in the standards, quite a large error is made.

Because of the large linear range in ICP-OES, bracketing calibration is not used for the purpose for which it was originally designed. Instead, it is used for analyses which require the best possible accuracy and excellent reproducibility. Examples are noble metals analysis and stoichiometric determinations. In these types of analyses, the total error must be as low as possible. Typically, one aims at a tolerance 0.1 %, and preferably below this figure. The bracketing calibration is used here to calibrate immediately before and after the measurement of the sample. Thus, small fluctuations in the complete system are cancelled most effectively. As a rule, the measurement of the sample (and with that the two matching calibration solutions) is repeated several times and averaged.

5.2.3 Examination of the Calibration Data

At the end of the calibration procedure, the calibration should be critically evaluated. As an example, a measurement protocol of the calibration, with comments, is shown in Fig. 124. (Note: As calibration techniques differ markedly from each other, the following remarks refer only to external calibration).

```
Method Name: Main components

Method Description: Dilution factor 1 : 5
```

Analyte	Calibration Equation	Processing	View
Ca 317.933	Lin, Calc Int	Peak Area	Axial
Cl 725.670	Lin, Calc Int	Peak Area	Axial
K 766.490	Lin, Calc Int	Peak Area	Radial
Mg 285.213	Lin, Calc Int	Peak Area	Axial
Na 589.592	Lin, Calc Int	Peak Area	Radial
Zn 206.200	Lin, Calc Int	Peak Area	Axial

===

Sample ID: Blank   Date Collected: 14.11.00

Analyte	Intensity	Std.Dev.	RSD	Conc.	Units
Ca 317.933	-775.7	96.73	12.47%	[0.00]	mg/L
Cl 725.670	12569.3	170.81	1.36%	[0.00]	mg/L
K 766.490	13741.7	131.26	0.96%	[0.00]	mg/L
Mg 285.213	872.6	75.30	8.63%	[0.00]	mg/L
Na 589.592	16178.3	98.63	0.61%	[0.00]	mg/L
Zn 206.200	117.5	21.95	18.68%	[0.00]	mg/L

Should be zero   **Should be infinitely large**

===

Sample ID: Standard   Date Collected: 14.11.00

Analyte	Intensity	Std.Dev.	RSD	Conc.	Units
Ca 317.933	12708251.4	104543.06	0.82%	[1]	mg/L
Cl 725.670	27428.7	313.16	1.14%	[1]	mg/L
K 766.490	6860628.8	22211.70	0.32%	[10]	mg/L
Mg 285.213	47957070.8	51672.63	0.11%	[5]	mg/L
Na 589.592	22180549.5	127669.77	0.58%	[50]	mg/L
Zn 206.200	188533.7	691.84	0.37%	[0.05]	mg/L

Should yield similar   **Should be as small as**
intensities as in   **possible, depending**
previous analyses   **on the spectrometer**

Fig. 124: A measurement protocol of a calibration procedure with comments. To start with, the readings for the blank are inspected, where the intensities should be close to 0 (zero). Obviously in the example shown, there is contamination of the blank for Cl, K, and Na since high intensities are measured and the reproducibility is very good at the same time. This behavior is typical of contaminations in the blank solution. High intensities are also found in the blank for Mg, but the RSD is quite large. This indicates that there are traces of Mg left in the sample introduction system, which are washed out by the blank solution. If such a contamination is observed, the calibration process should be interrupted and the causes of the deviations should be looked for. If a value close to "0" is expected, then the "0" must always be seen in relation to the sensitivity. If for example 1 mg/L of Mg yields an intensity of about 48 million counts per second, 873 for the blank reading can be regarded a very close to zero. However, if 1 mg/L Cl yields 27 429 c/s, then 12 569 c/s for the blank is excessive. On the other hand, roughly the same blank intensity for K is just acceptable in the view of the high intensity for the standard, about 69×10^6 c/s for 10 mg/L). With respect to the standard, one should check whether the sensitivity would remain roughly the same, so one should have an idea of which intensities to expect for the calibration standard from previous measurements. The reproducibility (RSD), at least of the highest standard of calibration, should be as good as the instrument is capable of. In this case, an RSD of about 0.5 % would be expected. When judging statistical data, one should bear in mind that, especially for a small number of repeats, these statistical indicators themselves have a large spread.

The reproducibility of the measurement of the calibration solution should be in the order of 0.1 to 3 % (depending on the spectrometer used, the sample introduction system, the processing technique and the matrix). The intensities found for the standards should be in the usual range. If spectra are displayed during the analysis, it should be checked whether the processing is performed at the peak center.

When measuring a blank solution, no peak should be seen in the window of the spectral display. Instead, a spectral range dominated by noise is to be expected. In the case of a structured background, obviously this structure must be seen. If no spectra are displayed during the analysis, the intensity of a blank should be near zero and, provided that a background correction is performed and there are no structures at the location where the intensity is measured for the blank, the precision should have high numerical value (theoretically infinite). If this is not the case, the measurement should be canceled and the cause for the high reading should be looked for. The most likely causes are contamination of the blank solution or a delayed rinse-out behavior ("memory-effect"). By repeatedly measuring the same blank solution, the important cause can be found. If the signal for the blank solution also remains constant after a longer time, it can be assumed that there is a contamination of the blank solution. If however the intensity decreases, then this is an indication of carryover where the sample introduction system is washed out slowly. In this case, the rinse time must be increased between the measurements. In the case of carryover, the use of lower concentrations (by diluting all the solutions) should be considered. A negative intensity with good reproducibility shows that the background correction was not set correctly.

5.3 Quality Assurance

Large numbers of analytical results are generated in the modern analytic laboratory in order to have a basis for economic and social decisions. For example, the specifications of (chemical) products are guaranteed or the harmlessness of an environmentally relevant emission is declared according to the analytical values. The assessment of the results obtained in the analytical laboratory will also have many material and immaterial consequences. With this in mind, the damage which a false result may cause cannot be overestimated.

However, the very fact that modern analytical instruments produce such large amounts of data can have the consequence that an individual result is checked less carefully, and the general tendency to cut costs under any circumstances makes this even more likely. Between this trend and the increasing importance of the analytical results the gap widens continually.

To cope with this situation, the principles of quality assurance were developed. The strategies used for analytical quality control (QC) originate from the quality supervision of mechanical series production [349]. Here the diameter of screws, for example, is checked, while in instrumental analysis a concentration is monitored. However, there is an immediate obvious problem in transplanting this quality control strategy. In series production control, the samples should all have the same predetermined dimensions to match the specification. However, in the supervision of the "production of results", this is impossible, because the true concentration in the sample is unknown. As a remedy for this, quality control samples with known concentrations are put into the series in regular intervals. If the expected concentrations are not found, this indicates a possible analysis error. The quality control sample should be a sample solution prepared independently from the calibration solutions. Standard reference materials are ideal for this purpose.

Validation and quality assurance are frequently confused. During the validation of a method, the analytical method developed is confirmed as suitable for a special application, while, with the help of quality control, errors during the routine measurement are recognized and corrected if possible.

A typical routine analysis run starts with the calibration and its evaluation. If the calibration function is found to be acceptable according to the (statistical) criteria discussed above, it is checked for accuracy by confirming that the concentrations of quality control samples are within tolerable limits. If these are also acceptable, then the samples are finally measured. After a set number of samples, this is then checked again by measuring the quality control samples. Only if it is confirmed that the calibration function is still valid will the next set of samples be measured. The measurement of quality control samples is repeated at regular intervals and at the conclusion of the analysis of the series of samples.

The principles expressed here were practiced by many users also before the introduction of standardized quality control measures. The noteworthy difference is primarily that the "experienced" analyst judged the quality of the measurements by "feel", while with quality control measures unbiased criteria are set and documented. The advantage of the "old" procedure consists in the fact that a reaction could be quite fast and flexible, the essential disadvantage being that it is strongly dependent on the person and the laboratory conditions. The standardized quality judgment is objective and compliant with regulations, but it is fairly labor-intensive.

As a rule, judgment is based on Shewhart quality control charts (Fig. 125). The most important parameters that will typically be registered are mean value, blank reading, recovery, and span widths. Details of the use of QC charts can be found in the literature [350].

In a QC control chart, the limits which describe the quality of the analysis are defined. In a preliminary period (normally 20 measurements), the parameters (e.g. mean value) of a reference material are measured under routine conditions. The standard deviation s of the mean values during the pre-period is then determined. The evaluation of the quality of the analysis is guided by two types of limits.

- Warning limits (± 2 s)
- Control limits (± 3 s)

The analysis is regarded as under control if the concentrations of the quality control sample are within the control limits. Results of quality control samples which are outside the control limits require an immediate reaction to remedy this out-of-control situation. A single value outside the warning limits is still acceptable if subsequent values are again within the warning range. However, if there is a trend such that seven results change in the same direction this will also be considered an out-of-control situation even if they are all within the warning limits. In exceptional cases, this can mean too slow a reaction time [351]. The use of QC control charts is a dynamic process because, when a chart is completed, all the results are used to calculate new warning and control limits.

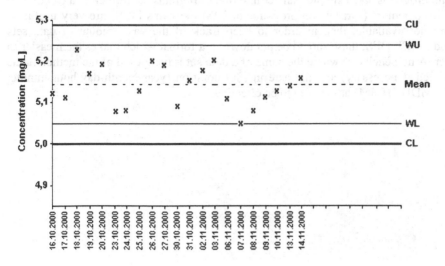

Fig. 125: Example of a mean QC control chart. (CU: upper control limit, WU: upper warning limit, Mean: mean average value, WL: lower warning limit and CL: lower control limit)

5.4   Software and Data Processing

The control of a complex system such as an ICP emission spectrometer requires efficient, well-designed and intuitively usable software. Meaningful default values and extensive help, such as wavelength tables which can be called up directly from the graphic display mode, simplify the operation. In some instruments, the software takes complete control of the spectrometer hardware so that a method can always be executed with the same optimized parameters. The importance of the software has greatly increased, and the claim made by many users, "The software is the instrument", becomes ever more realistic.

The sophisticated software consumes more and more computing power. Meanwhile, the enormous hard disk space allows all spectra taken during the analysis to be stored. This in turn makes an evaluation of the data possible after the analysis. If unfavorable processing parameters are found (e.g. faulty background correction), these processing parameters can easily be modified. In some software packages, one can recalculate the analysis with these changed parameters and with the stored spectra without having to remeasure the samples. This post-analysis evaluation and reprocessing of the spectra is a great help with difficult samples where unexpected spectral interference appears on the signal or on the background correction points.

Another great help for the assessment of the analytical quality is the representation of the spectra during the analysis for visual control of the accuracy of the results. Spectral interference which was overlooked or background correction points set at unfavorable positions can be recognized during this stage in order to correct the processing parameters.

The software or associated utilities should also help in archiving of the data. Integrated data bank programs and a simple transfer into other programs for further data calculation or data management (spreadsheet programs or LIMS systems [352]) are very useful. If these are not available, then in order to keep track of the vast amount of data sets produced in ICP-OES, these should be put down in a format which makes them easier to find later. A nomenclature where the name of a data set is composed of the method name (abbreviated if necessary) and the date in ISO notation (year-month-day-hour-minute: YYYYMMDDHHMM) has proved to be effective.

6 Trouble-shooting and Maintenance

Unfortunately, the instrument will sometimes just not work as expected despite all efforts and safety precautions. In this book, it is not reasonable to expect specific tips for all cases since these are quite different for the various spectrometers available on the market. Therefore, only generally valid statements are to be found in this chapter. Nevertheless, these should help in making the availability of the instrument as near as possible to 100 %.

An important tool in achieving good availability is an up-to-date logbook. Of course, this will not prevent malfunctions of the instrument, but a well-kept logbook helps to identify the error source faster when problems appear. In the logbook, every user writes down the date and time of the use of the instrument, the application, special observations, and actions taken for adjustment, care and maintenance.

As long as your instrument does what you expect of it, you should spend a little time and effort to get a feeling (and possibly data) for what makes a correctly operating ICP emission spectrometer:

- How long does the initialization take?
- What sounds do you hear at every stage of operation?
- How is the torch attached into relation to the induction coil?
- Where exactly does the plasma burn, how big is it, how bright?
- How even is the sample flow through the peristaltic pump?
- What does the aerosol look like (inside the nebulizer chamber or with the nebulizer not attached)?
- How warm is the RF generator or the torch compartment?
- What sensitivity and reproducibility can you expect?

Nevertheless, something may not run well. The most frequent failure situations are listed below:

- The plasma does not ignite
- The plasma flickers
- The plasma extinguishes
- No signal
- The sensitivity is significantly lower than usual
- The sensitivity changes markedly (drift)
- The short-term stability is inadequate.

Hints for trouble-shooting and what to do next are given below in separate boxes under the respective keyword. In the author's job as a trouble-shooter, suggestions for checking the simplest details were sometimes found to be quite helpful. These and more advanced tips are mentioned here.

✖   **The plasma doesn't ignite or extinguishes**

- Is enough argon gas available (sufficient pressure)?
- Was the sample introduction system purged thoroughly with argon to displace air before the igniting?
- Are the tubes leak-tight? A small leak will cause the plasma to extinguish with a rattling or hissing noise just after igniting.
- Is an ignition spark visible?
- Are the electric contacts oxidized?
- Is there a high load of dissolved matter which is not tolerated by the RF generator?
- Are organic solvents aspirated? – If yes, see Chapt. 7.3.7. "Organic solvents"

✖   **No signal**

- Is the peristaltic pump tube put in correctly? (Check the pump tension and direction of rotation.)
- Carry out the sodium bullet test. Aspirate a solution containing ca. 1 g/L Na and watch the plasma. After about a minute, the plasma should be colored orange-yellow. The analyte channel should be clearly visible in the center of the induction coil. If this is not the case, the following problems may exist:
    - The nebulizer does not nebulize
    - The pump does not transport any liquid
    - The injector has clogged.

The yellow bullet at the beginning of the analyte channel reveals the initial radiation zone. For radial viewing, this should end about 2 to 3 mm above the induction coil. For axial viewing, the plasma is frequently operated with a lower nebulizer gas flow. Then the yellow bullet ends within the induction coil. This is sometimes difficult to recognize.

If the bullet is too low, the nebulizer gas flow should be set higher, and *vice versa*.

You can do this test also with yttrium solution. Take the same concentration of Y (1 g/l). A red area indicates the emission from atomic transitions. These characterize the emission in the initial radiation zone and in the tail plume. The normal analytical zone appears as a blue zone of the analyte channel. Its emission is characterized by ionic transitions.

- Are the peaks at the correct wavelength position? If no, do a wavelength calibration of the spectrometer.

✗   **The sensitivity is significantly lower than usual**

- Is the loss of intensity the same for all wavelengths? Fogging of the window protecting the optics from the plasma will affect the transparency in the UV range particularly strongly while sometimes no change is observed in the visible range.
- When using spectrometers that can view in both directions (radially and axially), is the signal depression the same for the two viewing positions? This could be due either to poor adjustment of the torch with respect to the optics or to different extents of fogging of the windows from the plasma to the optics.
- Are the peaks at the correct wavelength position? If not, do a wavelength calibration of the spectrometer.

✗   **The sensitivity changes markedly (drift)**

Changes within the range of about 10–20 % are normal. The variable quality of the tubing of the peristaltic pump is frequently an essential factor. If the deviations are well outside this range, the causes should be clarified and rectified. (See also the notes in the previous box "The sensitivity is significantly lower than usual".)

Mermet suggests a procedure for finding the causes of drift [353] with the help of a set of emission lines. For drift diagnosis, he suggests the following lines: Ba 455, Zn 206, Ar 404, Mg 280 and Mg 285. With these lines, change in the energy transfer into the plasma may be diagnosed, resulting from changes in

- RF power
- Nebulizer gas flow
- Sample aspiration rate (nebulizer, pump).

Ba 455 does not react strongly to changes in the excitation conditions, since the sum of the ionization and excitation energy is small. The opposite is true for Zn 206, which reacts very strongly to changes in the excitation temperature. Both emission lines react similarly to changes in the nebulizer efficiency. Ar 404 follows changes in the excitation conditions. Mg I at 285 nm and Mg II at 280 nm represent the degree of the atomization or ionization and reflect changes in the RF power and the nebulizer gas flow, particularly in the case of a non-robust plasma ($I_{Mg\,I} / I_{Mg\,II} < 6$).

The suggestions for drift diagnostics are complemented by further diagnostic tools for precision, accuracy, stability, interferences, selectivity and limits of detection [354].

Table 10: Diagnostic aids according to Mermet [353]

Measuring what at which line	*Information on*
Ba II 233 nm FWHH	Resolution in the UV
Ba II 455 nm FWHH	Resolution in the visible range
Mg I 285 nm/mg II 280 nm	Atomization / ionization
Zn II 206 nm / Ba II 233 nm	Excitation
Ar I 404 nm	Absorption
Background 400 nm / background 200 nm	Poor transparency or reflectivity
Background 235 nm / background 236 nm	Poor transparency of fiber optics
Signal / background Mg I 285 nm	Clogged nebulizer
All wavelengths	Clogged nebulizer
RSD Mg I 285 nm	Clogged nebulizer
RSD Ar 404 nm	Drift / stability
RSD Zn 206 nm	Drift / stability
RSD Ba II 455 nm	Drift / stability
Standard deviation background 190 nm	Noise of background
Standard deviation without plasma	Detector noise
Ba II 233 nm limit of detection	Complete system
Zn II 206 nm limit of detection	Complete system

As a rule, a combination of the different bits of information gives a good idea of the malfunction of an instrument component. Salin alternatively suggests the thorough analysis of argon and hydrogen emission lines and the signal/background ratio of the blank spectrum as an indication of problems of the RF generator or the sample introduction system [355].

⚒ Inadequate short-term stability

The reproducibility/precision (RSD) is unsatisfactory (optimally it should be about 0.5 % depending on used spectrometer but not more than 3 %, provided that the concentration is significantly above the limit of detection). One should also consider that high matrix concentrations could be responsible for poor short-term stability. Is the reproducibility of a diluted solution satisfactory? If this is the case, the conditions must be optimized further or the samples must be diluted.

There are a number of possible instrument-related reasons for bad precision, which can be summarized in three groups:

Changes in
- The emission
- The optics
- The detector.

The temperature, which is high enough for the emission of the plasma, is influenced by different factors that are partially interdependent:
- Nebulizer gas flow
- RF power
- Mass introduced into the plasma. This depends on:
  - Concentration of the dissolved solids
  - Peristaltic pump rate
  - Nebulizer efficiency
  - Nebulizer chamber behavior
- Auxiliary gas flow
- Plasma gas flow
- Erratic drainage of waste
- Pressure changes caused by the exhaust system.

Poor reproducibility can also be caused by the optics. The most important error sources for this are:
- Absence of adjustment of the optical axis on the analyte channel
- Insufficient wavelength stability.

Finally, the noise of the detector can cause fluctuations of the measured signal.

The most frequent reason for non-reproducible measurements is a clogged nebulizer, which is fortunately simple to repair for most types of nebulizers. Therefore, you should check the nebulizer by visual inspection of the orifices:

a. Protect electronic and mechanical parts around the nebulizer with disposable towels, kitchen wipes or something similar and pump water (not the acidified rinse solution).

Does water arrive at the nebulizer after a little time?

If the answer is no, either the orifice or the sample supply tube is totally blocked (exchange or clean these respectively), the supply tubes have loosened, or the peristaltic pump does not transport the liquid properly (if the pump tube is OK, check the tension of the peristaltic pump).

b. Turn on the nebulizer gas. A fine aerosol flow should now be formed. This can be difficult to see. If necessary, spray against a kitchen towel or similar and check if the spot opposite of the argon orifice becomes moist.

If this does not happen, the argon orifice is blocked. Replace the orifice if possible or clean it. With some demountable nebulizers, the orifices may not be positioned correctly with respect to each other.

If solutions with high concentrations of dissolved substances are aspirated, they may crystallize or dry up at the tip of the injector. Organic solvents tend build up carbon deposits at the tip of the injector. In all cases, the injector must be cleaned. It should be adjusted in such a way that the distance to the plasma is increased by some (fractions of) millimeters. Alternatively, or in addition, the auxiliary gas flow should be increased. Both measures have an influence on the best (radial) viewing zone. This must be checked and if necessary re-adjusted.

In order to facilitate further trouble-shooting, solutions containing excessively high concentrations of dissolved solids should be excluded. Therefore, the following tests should be performed with diluted aqueous solutions. If the measurements are still fluctuating, the nebulizer and the nebulizer chamber and as a last resort the pump (see below) should be checked. If the poor reproducibility persist, check the mass flow controller or the valve for the nebulizer gas flow, and if necessary replace the mass-flow controller or valve with an external regulator for the time of the test.

Although during the normal mode of operation (regular inspection of the tube material) the pump tube should be working properly, you should carry out on-the-spot checks on the pump tube material for its usability and the peristaltic pump for its tension and even rotation:

- Are all connections between the tubes OK?
- Does the waste drain away evenly?

If this is not the case, you should replace the tube for the pumped waste. If the instrument has a freely draining waste, check, whether there are any blockages in the tubes or whether the waste tube is not squeezed flat somewhere. If a loop is provided for the waste tube, take care that the diameter of the loop is large enough (refer to the instrument manual) and that the loop is lined up vertically. The slope of the tube also should descend steadily and not have any upward bends, since otherwise a siphon effect can cause pressure changes in the nebulizer chamber, which affect the plasma for short periods, causing a significantly different signal, see Chapt. 2.4.2.4 "Waste from the Nebulizer Chamber".

If the surface of the nebulizer chamber is too smooth, this can also be a cause of poor reproducibility. The solution impacted on the nebulizer chamber walls should drain away evenly. A certain roughness of the surface of the nebulizer chamber is essential for this. If the surface is too smooth, larger water drops form, which disturb the aerodynamics in the chamber, see Chapt. 2.4.2.2 "Materials".

If a smooth surface of the nebulizer chamber should be responsible for poor short-term stability, the addition of a surfactant (e.g. Triton X) will yield an improvement. In this case, the surface of the nebulizer chamber should be etched. Depending on the material of the nebulizer chamber, this is done with different reagents. Glass or quartz is treated with a strongly diluted hydrofluoric acid (diluted about 1 : 40 to 1 : 100). Highly concentrated oxidizing acids [e.g. 50 % nitric acid (HNO_3)] attack the surface of Ryton, a material that is frequently used in sample introduction systems, and will lead to improved run-off properties. Make sure to constantly check the course of the etching process in order not to inadvertently destroy the material of the nebulizer chamber. When etching glass or quartz the progress can be generally observed visually: When nebulizing HF, at first drops form on the surface, and after a short time these will spread out and form a liquid film. The process should then be immediately discontinued and the nebulizer chamber should be thoroughly washed with water.

Is the gas supply line for argon and particularly for the nebulizer gas leak tight all the way to the injector? If necessary, spray a little soap solution onto the connections of the gas tubes. Next check the auxiliary gas line (and the valve) and then the plasma gas line (also the valve) and the regulators which reduce the pressure to the operating pressure, and finally the entire argon gas supply line.

Check whether the performance of the exhaust system is adequate. Take care that a small gap remains for extra air to enter between exhaust hood and the exhaust pipe of the instrument. There are cases known where stationary waves build up in the exhaust venting system, which adversely affect the plasma stability.

Verify the stability of the RF generator power by observing the short- and long-term stability of an argon emission line with no sample or liquid or air introduced and without nebulizer gas. The reproducibility should be less than 0.5 % and preferably lower, depending on instrument used. If this cannot be attained, it could indicate that the generator tube should be replaced.

If possible, you may want to record the wavelength position of the argon emission line at every measurement during this test in order to get an idea of the wavelength stability of the instrument. Use all possibilities for wavelength stabilization offered by the spectrometer. In particular, ensure that the spectrometer is calibrated with respect to the wavelength axis such that the center of the peak coincides with the wavelength expected.

The sometimes difficult step of setting the optical axis to coincide with the analyte channel should be carried out now at the latest. If this step is simple (sometimes it is supported by the software of the instrument), it is advisable to perform it at a much earlier stage of the proceedings.

If the search for error sources has not yet been successful, you should inspect the mass flow controller or the needle valve controlling the nebulizer gas. If necessary, replace the internal mass flow controller with an external one or by a needle valve. Note that a good needle valve is certainly better than a poor mass flow controller and should be used for test purposes.

A very rare cause (depending on the spectrometer) is an increase in the detector noise. If the detector has to be replaced, this must be carried out by customer services.

To avoid the failures described above, you should check your instrument at regular intervals, which will vary according to the instrument, the application(s) and the daily running time. You should inspect the instrument and its components, but under no circumstances dismantle them merely for a visual inspection. Remember the motto: "Never change a winning team!" – or, modified for instrumental analysis: "Never change a running system"!

The attention is principally focused on the sample introduction system, particularly the nebulizer. Analytical tests (sensitivity, reproducibility) will give sufficient information on the state of the nebulizer. The injector should not have any deposits, but if any are in fact present, clean the injector; when working with aqueous samples, treat it with solvents in the order: water, dilute acid, concentrated (oxidizing) acid, aqua regia. The fact that there are deposits should prompt you to increase the injector-plasma distance as a possible remedy. The distance between the injector tip and the lower edge of the induction coil is about 1 mm for many instruments. If deposits are present, this distance should be increased by several tenths of a millimeter. Alternatively, the auxiliary gas flow can be increased.

Demountable torches and sample introduction systems generally contain a number of O-rings for sealing. Do not disassemble your torch or sample introduction system just to check whether these still are all right because you may destroy them in the process of checking. Inspect the O-rings only if you are compelled to remove the corresponding component. In regular intervals, you should check that there is no backfill in the nebulizer chamber.

If deposits on the quartz parts of the torch are noticed, the torch should be cleaned. Metal films are particularly critical, because they may drain energy from the RF generator. However, even boiling the contaminated quartz with aqua regia will often not remove the deposits, as these may have penetrated some molecular layers deep into the quartz. Only a very short dip in hydrofluoric acid then helps, immediately followed by a vigorous rinse with water. For this procedure, the recommended security measures (gloves, aprons etc. to protect any exposed skin from spillages, safety goggles, exhaust hood) must be strictly adhered to! This very aggressive treatment attacks the quartz of the torch and should therefore only be done as a very last resort. The coating will be removed, but the surface becomes rougher and the formation of deposits will be accelerated.

The torch becomes rough on the inside as part of normal wear, but the wear is speeded up by high concentrations of alkali metals. Occasionally (preferably before you start up the plasma, to avoid burning yourself), you may want to feel the roughness by touching the inside of the outer tube with the little finger. (Make sure you will not carry any sweat onto the torch this way because the salt in it will speed up the devitrification of the torch, see last paragraph in Chapt. 2.1.2 "Plasma Torch".) If the quartz inside is very rough, the torch should be replaced.

The test for transmission of light from the plasma to the optics is usually performed by regularly recording the sensitivity of selected analytical lines in the vacuum-UV (at or under 190 nm) in a typical wavelength range (around 250 nm) and in the visible range (400 to 800 nm), if these are relevant. Fogging of the quartz window or fiber optics will first be noticed in the deep UV range. Ratios of background intensities (see Table 10) can be used as an alternative indicator of decreasing transparency of the transfer optics or fiber optics. If necessary, clean or exchange the corresponding quartz windows or fiber optics.

Usually there are air filters, which often are on the back of the instrument. Check whether a sufficient amount of air can pass still through them. Otherwise, the tube of your generator could fail through overheating. If necessary, clean the filters, preferably in a dry state. Under no circumstance, install a wet or moist filter!

If you cool with a recirculating cooler, you should make the following checks also:
- Water level
- Algae or fungus growth
- Permeability of the filters to air and water.

7 Applications

7.1 General Notes

First and foremost, avoid the use of glass, particularly when working in the trace range and especially in the extreme trace range. The absorption and subsequent desorption presents many an unpleasant surprise. In addition, some components of sodium borosilicate glass (Na, B, Si) may be leached out and may influence the determination of low concentrations of these elements. Volumetric flasks made of glass have caused quite a number of false results.

The fewer steps there are in the analytical procedure, the fewer possibilities for errors exist. Over years, the author has had good results when filling the samples directly into the containers for the auto-sampler, particularly when doing trace analysis. This suggestion contradicts the procedure, which is still taught, to use a volumetric flask to make up liquid samples to a particular volume. The contamination risk more than outweighs the potential volumetric error. The possible volume error is in the same range as the statistical error of the subsequent determination, which is in the order of up to a few percent, and is therefore negligible. In contrast, the error due to contamination or absorption can be up to several orders of magnitude! Moreover, the smallest dilution error is made if solutions are weighed out for making up to a volume (if one just takes a note of the weight it is also much faster) and also if the diluted solutions are weighed.

When using volumetric flasks, experience shows that sometimes the mixing step will be carried out inefficiently, especially if the air cushion is small. In order to obtain a good distribution of the sample in the final solution, it has to be shaken long and vigorously.

All solutions that are to be measured, which include the calibration solutions and samples, should contain a medium that ensures stability. As a rule, these solutions are acidic. Now and then one must depart from this practice since the sample solution is for example alkaline. Then the calibration solutions must be prepared in exactly the same medium for optimal matrix matching. However, in alkaline solutions many metals have limited solubility with respect to concentration and time. Therefore, the sample solutions should be measured as soon as possible after their preparation [356]. Of course, this also applies to the calibration solutions, which should also be prepared directly before use.

As stated, acidic solutions can generally be kept for a longer time. The concentration of the acid (as a rule here: the more the better) frequently decides the stability of the analytes in the solution. If the concentration is not in the range of the solubility product of the ions involved, concentrated solutions are more stable than strongly diluted ones. The stability of solutions, which for example often contain a concentration of analytes in the mg/L range, is in the region of weeks, sometimes even months. In the very low µg/L range, absorption on the container walls is already noticeable after a few days.

It is generally true to say that the higher the concentration, the larger is the analyte signal. There are exceptions, though. Particularly at a high salt load (e.g. 20 % NaCl) , non-spectral interference (predominantly excitation interference) is sometimes so severe that the sensitivity can decrease dramatically. The excitation interference by easily ionizable elements (e.g. alkali metals) occurs not only thermally but also by changing the electron equilibrium in the plasma [357]. In general, the presence of alkali metals causes an increase in the intensity of the analyte channel near the induction coil and a decrease further away from it [358].

On diluting, this effect of the signal depression is reduced, so that the unusual effect is observed that the signal increases with increasing dilution because of lower signal depression. At a certain point, the sensitivity no longer increases with increasing dilution, and after this the intensity of the analyte signal decreases due to the dilution effect. Therefore, the optimal concentration of the salt should be determined experimentally by checking the sensitivity as a function of the dilution factor.

Contamination is ubiquitous in the trace range. The lower the concentration to be determined, the more care must be taken at every step [359]. The reagents and the water [360, 361] for dilution must be extremely pure. Even then, contamination cannot be avoided completely. As an example, one should always anticipate contamination by Si, e.g. if the acid for digestion is kept in a glass bottle.

7.2   Comments on Selected Elements

As with every other technique, ICP-OES has analytes that can be analyzed particularly well. Mn is such an element, where one must make an effort to produce unsatisfactory results. The alkaline earth metals can be measured with an extremely good detection limits [362]. However, an advantage in one respect can be the reverse in another: The elements Ca and Mg are ubiquitous, so they typically yield measurable signals even in ultra-pure water.

Generally, it can be stated that the ionization in the plasma and the excitation to an emission decreases in the periodic system from left to right Degree of. The nonmetals need more energy. Therefore, for example, the most sensitive emission lines of the nonmetals are found in the very low vacuum-UV range (Fig. 126). As a rule of thumb, one could say that metals can be determined better than nonmetals with ICP-OES. Some specific features are listed in Table 11.

Table 11: Specific aspects of selected elements analyzed by ICP-OES

Al	Contamination risk in the trace range [363]! Ubiquitous element. As a highly concentrated matrix component, it causes molecular band structures in the lower UV range. Pretreatment with HF is recommended for complete digestion [364, 365].
As	In a dry aerosol (e.g. ultrasonic nebulizer), the sensitivity depends on the oxidation state. Not very sensitive.
B	Extremely high carryover [366]! The element is easily adsorbed on the walls of the sample introduction system and is washed out again only slowly. Addition of mannitol [367] or ammonia [368] reduces the memory-effect.
Ca	Very sensitive. Ubiquitous: a signal for the most sensitive lines is observed in clean blank solutions.
Cr	In a dry aerosol (e.g. ultrasonic nebulizer), the sensitivity depends on the oxidation step.
Hg	Not very sensitive in relation to the anticipated concentrations in most samples. High carryover risk!
I	Strong dependence on the oxidation state, the chemical bonding and the redox potential in the solution [369].
K	Small linear range. High susceptibility to excitation interference, particularly on axial viewing.
Mg	Very sensitive. Ubiquitous: a signal for the most sensitive lines is observed in clean blank solutions.
Na	Small linear range. High susceptibility to excitation interference particularly on axial viewing.
S	Strong dependence on the oxidation state and chemical bonding. Compounds that may form H_2S (due to the formation of this gas) or are volatile themselves have a higher sensitivity than ions such as sulfate [370].
Si	Volatile in presence of high HF concentrations, which causes losses from the solution (before the measurement) and/or higher sensitivity (due to the formation of gas). Si may volatilize in the nebulizer chamber in the presence of HF to form SiF_4. This effect can be blocked by addition of boric acid or tertiary amines [371]. Precipitation of silica in the lower pH range.
Sn	In oxidizing acids, a precipitate of stannous oxide forms. Not very sensitive.
Se	Not stable in HCl, since volatile. Not very sensitive.
Zn	Contamination risk! Ubiquitous element.

A number of nonmetals have very sensitive emission lines in the low vacuum-UV. This is exploited particularly in the determination of halogens by ICP-OES [372, 373].

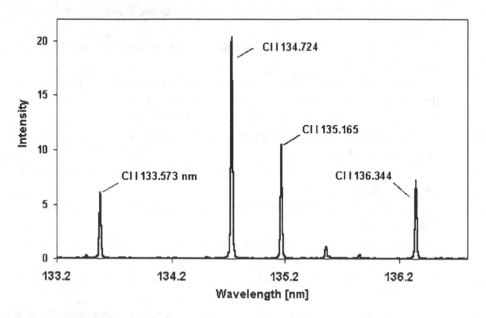

Fig. 126: Cl has some very sensitive atomic emission lines in the low vacuum-UV in addition to the insufficiently sensitive emission lines at 725.670 and 782.139 nm (source: Spectro Analytical instrument)

7.3 Selected Applications

As in other fields, there are no universal procedures or suggestions which always work in every application using ICP OES. Every application has its own particular requirements and weak areas. On top of this, every instrument is different. Nevertheless, there is a background of experience for a number of applications which at least provides general guidance. The following section will include some thoughts that may assist with method development for a particular application. Even though lists of wavelengths are included, you should check their suitability with your own spectrometer, since spectrometers differ greatly in a number of key qualities (e.g. resolution, utilizable wavelength range, generator and sample introduction system).

✂ **Classification of applications**

7.3.1 Environment
7.3.1.1 Drinking, ground and surface water
7.3.1.2 Waste water, leachates
7.3.1.3 Sludges
7.3.1.4 Soil, sediment
7.3.1.5 Airborne particles, fly ashes

7.3.2 Samples of biological origin
7.3.2.1 Plant and animal samples
7.3.2.2 Clinical and forensic materials
7.3.2.3 Food and animal feeds

7.3.3 Geological materials

7.3.4 Metallurgy
7.3.4.1 Steel and iron matrices
7.3.4.2 Non-ferrous metals
7.3.4.3 Noble metals
7.3.4.4 Special alloys

7.3.5 Material sciences
7.3.5.1 Semiconductors
7.3.5.2 Ceramics

7.3.6 Industrial applications
7.3.6.1 Chemicals and fertilizers
7.3.6.2 Galvanizing/electro-plating baths
7.3.6.3 Brines and salts
7.3.6.4 Cement, gypsum, calcium matrix
7.3.6.5 Glass
7.3.6.6 Other industrial applications

7.3.7 Organic solvents
7.3.7.1 Wear metals and contamination in oil
7.3.7.2 Additives
7.3.7.3 Tar
7.3.7.4 Edible oils

7.3.1 Environment

The prime goal of environmental analysis is the supervision of tolerance limits in a variety of "natural" matrices of extremely diverse composition. The analytical task is to measure concentrations, which should be as far as possible below the permitted concentrations. If this requirement is fulfilled, a relatively large error is typically accepted. If, however, a concentration is found to be near the limiting value or even above it, then the demand for accuracy and reproducibility greatly increases. These requirements should be kept in mind as early as the line selection stage. A wavelength may then be usable even if there is slight spectral interference if the measured concentration is well below the tolerance limit.

Other environmental analytical questions are (rarely) estimates of the overall mass discharged to or present in an environmental compartment, and (very much more rarely) research into the origins of pollution [374]. Much higher accuracy is required for the latter types of applications.

As a rule, for these types of applications, extracts of solid samples are used for sample preparation.

Extremely polluted environmental samples can have extremely variable compositions. Therefore, interference which is not normally important in one sample, may suddenly dominate in another. In order to safeguard the results in such cases, it is recommended to determine an element using several lines [375]. Particularly contract laboratories, which often have little information on the sample, are well advised to use this defensive strategy. On the other hand, there are applications which typically do not hold any surprises. An example is the supervision of e.g. the cooling water of a power plant, which is quite constant in its composition, so there is a far lower risk of unexpected interference.

7.3.1.1 Drinking, Ground and Surface Water

Legislators set tolerance limits for a number of pollutants in drinking water (for example, the European Community guideline on drinking water [376]). The toxicologically relevant elements have to be determined with a predetermined measurement error. The tolerance values in accordance with the guideline on the quality of water for human consumption are listed in Table 12. These are compared with the permitted limits of detection referred to in the method ISO 11885, which apply to ICP-OES analysis with radial viewing. Using axial viewing, as a rule, the limits of detection improve by around an order of magnitude. Furthermore, the method ISO 11885 permits gentle evaporation of the sample up to an enrichment factor of 10 [377]. Most of the elements can be determined using a pneumatic nebulizer [378, 379]. Others must be measured using an ultrasonic nebulizer for the purpose of enrichment [380, 381, 382, 383]. The evaporation stage takes place virtually "on-line" in the desolvation unit of the ultrasonic nebulizer.

Table 12 contains suitable analytical lines and alternatives.

Table 12: Comparison of the legal limits in drinking water in accordance with the European Community guideline on water quality, the limits of detection using radial viewing in accordance with ISO 11885 and suggested analytical lines as well as alternative lines

Element	Legal limit [μg/L]	Limit of detection [μg/L]	Wavelength [nm]	Alternative wavelength [nm]
Ag		20	328.068	338.289
Al	200	100	396.152	167.022
As	10	80	188.979	197.197
B		6 (10)*	249.773	249.678
Ba		2	455.403	233.527
Be		**	313.107	313.042
Ca		10	317.933	315.887
Cd	5	10	228.802	214.438
Co		10	228.616	
Cr	50	10	267.716	205.552
Cu	2000	10	327.393	324.754
Fe	200	20	259.940	238.204
K		**	766.490	
Li		2	670.781	460.286
Mg		1	285.213	279.079
Mn	50	2	257.610	293.306
Mo		30	202.030	204.598
Na		100	589.592	330.237
Ni	20	***	231.604	221.648
P		100	213.618	178.287
Pb	10	200	220.353	283.306
S (SO$_4^{2-}$)		500	182.037	180.669
Sb		100	206.833	217.581
Se		100	196.026	
Si		20	251.611	212.412
Sr		10	421.552	407.771
Ti		5	334.941	336.121
Tl			190.800	
V		10	292.402	
Zn	–	5	206.200	213.857

*   In ISO 11885, a limit of detection of 10 μ g/L is given for the wavelength at 249.773 nm and of 6 μg/L at 249.678 nm. Since the wavelength at 249.773 nm is clearly more sensitive, it is assumed that there is a misprint.
**   No value given.
***   No value given. In the earlier standard, DIN 38 406 E22, a value of 20 μg/L was given.

The ICP-OES determination of sulfate in drinking water can be carried out by determining sulfur and calculating for sulfate [384].

7.3.1.2 Waste water, Leachates

The analysis of waste water (and sludge) is one of the major environmental applications [385, 386, 387]. Concentrations of pollutants in waste water samples range from drinking water quality up to highly polluted samples such as residues from electro-plating baths.

Sample pretreatment steps differ depending on whether the dissolved content or total content is to be determined. For dissolved components, the sample is filtered immediately after sampling through a 0.45 μm filter. The first part of the sample (50–100 mL) is used to wash the filter and is discarded. For preservation purposes, 100 mL of sample are acidified with 0.5 mL nitric acids (or until the pH < 2). This is now the solution which can be measured. If the total content has to be determined, the sample is not filtered, but is acidified as described above for preservation.

For digestion, ISO 11885 suggests the addition of 0.5 mL nitric acid per 100 mL sample. The solution is the evaporated almost to dryness, taken up with 1 mL nitric acid and made up to 100 mL. In a determination of Sn, the digestion must be done with sulfuric acid and hydrogen peroxide [388].

Leachates which are not heavily polluted can be digested with the help of UV radiation. Hydrogen peroxide serves as an oxidation agent.

Table 13 gives examples of analytical lines for waste water. However, the best lines for a particular application can differ from these due to raised levels of pollution!

Table 13: Analytical lines for environmental samples

Element	Wavelength [nm]	Alternative wavelength [nm]	Further wavelength [nm]
Ag	328.068		
Al	396.152	167.022	308.215
As	188.979	197.197	193.696
B	249.773	249.678	208.959
Ba	233.527	455.403	
Be	313.107	234.861	313.042
Ca	315.887	317.933	422.673
Cd	228.802	214.438	226.502
Co	228.616		
Cr	267.716	205.552	283.563
Cu	327.396	324.754	
Fe	259.940	238.204	
K	766.490	769.896	
Mg	285.213	279.079	279.553
Mn	257.610		
Mo	202.030	204.598	
Na	589.592	330.237	588.995
Ni	231.604	221.648*	
P	213.618	178.287	214.428
Pb	220.353		
S	182.037		
Sb	217.581	206.833	
Se	196.026		
Si	251.611		
Sn	189.927		
Sr	407.771	421.552	
Ti	334.941	336.121	
Tl	190.801		
V	292.465*	292.402	
Zn	206.200	213.856	

*: Not listed in ISO 11885

7.3.1.3 Sludges

Sludge samples are in effect concentrates of wastewaters. For sludge samples, a distinction can be made between domestic or industrial origin according to the pollution load [389]. In view of the large pollution load range on the one hand and the similarity of some types of samples (e.g. waste water and sludges) on the other, methods developed for one application could sometimes be applicable for another. An example would be that a method for domestic sludges could generally be used for only slightly polluted waste waters. In addition, methods for industrial sludges might be applicable to waste waters of the same factory.

The use of the sludges as fertilizers in farming [390] has triggered an enormous need for supervisory analyses [391]. ICP-OES is very suitable analytical technique for this application [392, 393, 394].

Aqua regia extraction is usually used as the digestion procedure. The sample is milled and dried for 30 min at 105 °C. To 3 g of sample, 28 mL aqua regia are added and left at room temperature for several hours. Then it is boiled under reflux for 2 h. After filtration, the extract is made up to 100 mL, the final volume.

Table 13 contains analytical lines, which can usually also be used for the analysis of sludges.

7.3.1.4 Soil Samples, Sediments

The matrices of the soils can vary greatly (e.g. Si matrix: sandy soil, Al matrix: clay, Ca matrix: weathered lime rock), so that no uniform recommendation for the line selection can be given. Generally, the wavelengths listed in table 13 for waste waters and sludges are also a good first start for the selection of analytical lines for soil samples [395, 396, 397, 398, 399, 400].

For the analysis of pollutants, an aqua regia extract as described for sludges is typically performed [401, 402]. In the determination of principal ingredients and nutrients in soil samples for the purpose of recommending a fertilizer, an extraction is carried out with calcium acetate lactate (CAL).

7.3.1.5 Air-borne Particles, Fly Ashes

The use of ICP-OES for the analysis of dust samples, whether these are factory plant discharges or airborne substances, dates back to the time of the commercial introduction of this method of analysis [403].

Airborne particles are collected with the aid of sampling systems containing a cascade of filters [404, 405]. Depending on the amount of deposits, either the whole filters or parts of it are digested. A quantity of sample (125 mg) is digested in 25 mL of a nitric acid/perchloric acid/hydrofluoric acid mixture at normal pressure in PTFE vessels [406, 407, 408]. Working with perchloric acid presents a potential explosion risk, so that extreme care should be taken. The safety standards described in the literature must be strictly observed [409]. Without the use of perchloric acid, a residue remains which has to be filtered off [410]. Alternatively, a digestion can be carried out by fusing with lithium tetraborate or lithium metaborate. Treatment with aqua regia, as is frequently done with sludge and soil samples, will not dissolve the sample completely.

For fly ashes [411, 412], a similar digestion is used [413]. However, the amount of dissolved matter in the sample solutions is much higher, so that matrix matching is absolutely required [414]. Alternatively, a fusion digestion can also be used [415].

Table 13 gives a guide to the selection of analytical lines.

7.3.2 Samples of Biological Origin

This group mainly comprises samples from the food industry, but other materials with a similar matrix composition are also discussed. There are mainly two goals for the analysis. One is to monitor compliance with specifications in processed foods or diet additives. Here a high level of accuracy is generally required (the tolerance level must typically not exceed few per cent). Another aim is to prove that toxicologically relevant elements are not present in the samples. Here the goal is to document their absence, and limits of detection as low as possible are required. While the system is optimized for high stability and absence of interference in the first case, every effort is made to obtain maximize the sensitivity in the second case.

In addition, clinical and forensic material is included in this group because of the similarity in the matrix composition. ICP-OES is very well suited for the determination of the major and minor components and some traces [416]. The use of an ultrasonic nebulizer extends the working range to low concentrations [417].

The analytical wavelengths are summarized in Table 14, which can serve as a basis for method development.

7.3.2.1 Plant and Animal Samples

Most samples can be digested in nitric acid [418, 419], if necessary with addition of hydrogen peroxide [420], hydrochloric acid [421], sulfuric acid [422] or hydrofluoric acid [423, 424]. Fatty substances can only be brought into solution in this acid mixture by microwave-assisted digestion, which also offers advantages for other types of samples [425]. During the digestion, "carbon compounds" in the matrix are transformed into carbon dioxide and water. As a result, the matrix of the digested sample solution is comparatively simple. Matrix matching nevertheless makes sense, since the digestion acids have an influence on the sensitivity [426].

7.3.2.2 Clinical and Forensic Materials

The first applications were published as early as 1980 [427, 428]. It must be said that ICP-OES has not become an accepted technique for this type of application, although some papers have been published its use [429, 430, 431, 432, 433]. Classical methods or AAS predominate here, since generally only a small number elements are determined. Also, graphite furnace AAS has better detection limits for important elements. This technique also has the advantage that the digestion process takes place in the graphite furnace, so that sample pretreatment is much less than it would be for ICP-OES. For some elements, such as Al [434], Si [435], Be [436], Sr [437], and Ba [438], ICP-OES is preferable because it has better detection limits for these elements. Sample taking is a frequently underestimated contamination source [439].

For the digestion of blood sample, approx. 6 g sample is mixed with 15 mL HNO_3 (65 %, suprapur) and left for 30 minutes at room temperature. Then it is heated up to light boiling for one hour. After cooling down, 3 mL H_2O_2 are added and carefully warmed until gas evolves. This is held at room temperature for 15 minutes and then slowly heated to boiling and held for one hour at boiling temperature so that a part of the acid evaporates. The elements Ca, Cu, Fe, K, Mg, Mn, Na, S, Si and Zn can be determined well.

Serum can be analyzed directly after dilution (1 : 3 to 1 : 10) with 18 MOhm water [440, 441]. The higher the dilution factor, the lower is the potential for non-spectral interference affecting the results. Therefore high dilution factors (up to 1 : 100) are preferred, provided that this does not reduce the sensitivity too much [442]. The sample introduction system should be thoroughly rinsed with water when changing between acidic blood digestion, standard solutions and neutral dilutions of serum, to remove all acid remains from the nebulizer chamber. If acid remains are still present in the nebulizer chamber, proteins will precipitate. Likewise, when changing from serum to acidic solutions the sample introduction system should be rinsed thoroughly for the same reason. Some authors prefer an acid digestion for serum [443].

Urine samples are stabilized with 1 mL nitric acid and measured directly after dilution by the factor of between 1 : 5 [444], 1 : 10 [445] and 1 : 50 [446] or with the addition of the internal standard Y [447]. Heavy metals must be enriched before a determination with ICP-OES [448]. Since the composition can vary considerably, maximum dilution is advisable as far as the limits of quantification allow this.

7.3.2.3 Food and Animal Feeds

Food samples are either liquids or solids, and can be of vegetable or animal origin. Solid food samples are treated according to the procedure described in Sect. 7.3.2.1 "Plant and animal samples" [449, 450]. In addition, foods in liquid form such milk [451, 452, 453, 454] or fruit juices [455, 456] should be digested.

Table 14: Analytical lines for samples of biological origin

Element	Biological samples wavelength [nm]	Clinical samples wavelength [nm]	Alternative wavelength [nm]	Further wavelength [nm]
Al	396.152	396.152	167.022	237.313
As	197.197		188.979	193.696
B	249.773	249.773		
Ba	455.403			
Be	313.107		313.042	
Ca	317.933	393.366	396.847	422.673
Cd	228.802		226.502	
Co	228.616			
Cr	267.716		205.552	357.869
Cu	327.396	327.396	324.754	
Fe	259.940	259.940	238.204	234.329
K	766.490	766.490		
Li		670.781		
Mg	285.213	279.553	279.079	
Mn	257.610	257.610		
Mo	202.030	202.030		
Na	589.592	589.592	330.237	588.995
Ni	231.604		221.648	
P	213.618		214.914	178.287
Pb	220.353			
S	182.037	182.037	180.669	
Se	196.026	196.026		
Si	251.611	251.611		
Sr	407.771		421.552	
Ti	334.941	334.941		
Zn	206.200	206.200	213.856	202.551

There are also food samples which are mainly of synthetic origin, such as refreshment drinks. These also can be analyzed directly after dilution. Beer can also be analyzed with ICP-OES without previous digestion after diluting by a factor of at least 1 : 5. Matrix matching is inevitable for this application. Wine is analyzed after digestion with hydrogen peroxide [457]. Animal feeds typically must also be digested [458].

Edible oils can be analyzed after diluting in an organic solvent (compare Sect. 7.3.7.4 "Edible oils").

7.3.3 Geological Materials

Geological samples include a variety of very different matrices and applications. The digestion methods are as different as the materials, and are mainly acid digestions [459] and fusion digestions [460]. Even an attempt at a full discussion would beyond the scope of this book. An excellent documentation is given by Heinrichs and Herrmann [461]. An indication of the wavelength selection for rock samples is to be found in Table 15. However, the choice of analytical line must be made for each type of rock individually. A number of publications exist on this topic [462, 463, 464, 465, 466, 467, 468, 469], especially on the determination of rare earth elements in rocks [470, 471, 472, 473, 474, 475, 476].

Table 15: Analytical lines for geological samples

Element	Rocks wavelength [nm]	Alternative wavelength [nm]	Further wavelength [nm]
Al$_2$O$_3$	308.215	396.152	309.271
B	249.773	249.678	
Ba	455.403	233.527	
Be	313.107	313.042	
CaO	315.887	317.933	422.673
Cd	228.802	214.438	226.502
Ce	413.765	418.660	
Co	228.616	238.892	
Cr	267.716	205.552	283.563
Cu	324.754	327.393	224.700
Dy	353.170		
Er	337.271	369.265	
Eu	381.967		
Fe$_2$O$_3$	259.940	238.204	239.562
Gd	342.247	335.047	376.839
Hf	277.336	232.247	
Ho	345.600		
In	230.606		
K$_2$O	766.490	769.896	
La	379.478	333.749	408.672
Li	670.781	610.362	
Lu	291.139	261.542	
MgO	285.213	279.553	279.079
MnO	257.610	259.393	
Mo	202.030	281.615	

Table 15, continued

Na$_2$O	589.592	330.237	
Nb	316.340	309.418	295.088
Nd	406.109	401.225	430.358
Ni	231.604	221.647	
P$_2$O$_5$	213.618	178.283	214.914
Pb	220.353	217.000	
Pr	390.844	422.293	414.311
S	182.563	180.731	
SiO$_2$	251.611	212.412	288.158
Sm	359.260	356.827	373.920
Sn	189.927	235.485	283.998
Sr	407.771		
Tb	350.917		
Th	283.730	359.959	
TiO$_2$	334.940	336.121	
Tm	313.126		
U	367.007	385.958	409.014
V	290.882	292.465	292.402
Y	371.030		
Yb	369.419	328.937	
Zn	206.200	213.856	
Zr	343.823	349.621	

7.3.4 Metallurgy

The main application is the supervision of the composition of alloys. High precision, usually involving high concentrations, is required here. Also, in some application areas, purity determinations (particularly in noble metal analysis) are carried out. In this, all potential contaminations are determined and the limits of detection must be taken into account. The sum of the found or assumed concentrations subtracted from 100 % yields the maximum purity that can be specified. For these applications, limits of detection as low as possible are essential.

7.3.4.1 Steel and Iron Matrices

Even a small change in the concentration of an alloying component can have a large impact on the properties of steel. Therefore, these types of analyses require a high degree of precision. For calibration, digested reference materials are often used. In order to keep the error as low as possible, the results are normalized, i.e. sum of all measured components is 100 % [477]. The use of an internal standard also aims at guaranteeing the highest possible accuracy [478].

Sample preparation of steels depends largely on the alloying additions. As a rule, nitric acid and sulfuric acid form the basis of the acid mixture, which is sometimes complemented by hydrofluoric and phosphoric acids [479]. The presence of phosphoric acid prevents the loss of boric acid by evaporation along with water vapor [480]. The most sensitive analytical lines of the element B are subject to spectral interference by Fe lines in steel analysis. Therefore, either this element is measured at a relatively insensitive line in the vacuum-UV, the matrix is removed [481, 482, 483], or multivariate regression techniques are applied [484].

A very thorough documentation is given by Gillum and Vail on the interference of the analytes Al, As, B, Ca, Ce, Co, Cr, Cu, Hf, La, Mg, Mn, Mo, Nb, Ni, P, Si, Sn, Ta, Ti, V, W, Zn, and Zr on each other and by the matrix element Fe [485]. The spectra were recorded with a spectrometer of 8 pm resolution in the wavelength range below 400 nm and a resolution of 15 pm above 400 nm. The user gets an excellent impression of the possible interference effects from the tables summarizing the examined elements and the display of the spectra of the most important wavelengths inspected.

Table 16: Analytical lines for steel and iron-based alloys

Element	Wavelength [nm]	Alternative wavelength [nm]
Al	396.152	308.215
B	182.598	249.773
Ca	393.366	317.933
Cd	226.502	
Ce	456.236	413.765
Co	228.616*	
Cr	267.716	
Cu	327.396	324.754
Hf	264.141	
La	404.291	333.749
Mg	279.553	285.213
Mn	257.610	403.075
Mo	202.030	281.615
Nb	319.498	313.079
Ni	231.604	221.647
P	178.283	213.617
Si	251.611	288.158
Sn	189.989	
Ta	263.558	301.254
Te	214.284	
Ti	337.280	334.940
V	309.311	311.071
W	207.911	
Zn	206.200	213.856
Zr	343.823	339.197

*: Ti-interference

7.3.4.2 Non-ferrous Metals

Aluminum is dissolved in sodium hydroxide solution and then acidified with nitric acid. The aluminum matrix causes a strongly structured background in the range 190–212 nm.

Copper may be dissolved by 50 % nitric acid [486] or by the addition of 5 mL hydrochloric acid and 15 mL nitric to 2 g of sample. If necessary, the reaction mixture must be heated [487].

Bronze is brought into solution with aqua regia [488].

A zirconium matrix is very line rich. Therefore, the selection of suitable wavelengths is particularly difficult. Sometimes corrections of the inevitable interference effects is necessary [489]. The digestion is done with a mixture of hydrofluoric and nitric acids. An alternative is matrix separation [490].

Possible analytical lines are summarized in Table 17.

Table 17: Analytical lines for selected non-ferrous metals and their alloys

Elem.	Al matrix wavelength [nm]	Bullet lead wavelength [nm]	Pure copper wavelength [nm]	Tin bronze wavelength [nm]	Mg matrix wavelength [nm]	Zr alloy wavelength [nm]
Ag		338.289	328.068	338.289		
Al	237.313		396.152		396.152	308.215
As		193.696	189.042	193.696		
Au		267.595		242.795		
B	249.773					182.589
Be	313.042		313.107			
Ca	393.366		393.366			
Cd		214.438	214.438	226.502		214.438
Cr	267.716			267.716		205.552
Cu	327.393	324.754	221.458	224.700	324.754	224.700
Fe	259.940	238.204	259.940	259.940	238.204	239.562
Ga	417.206					
Hf						273.876
Mg	279.553		279.553			279.553
Mn	259.373		257.610	257.610	257.610	293.930
Nb						309.417
Ni	221.648	231.604	231.604	231.604	231.604	221.647
P	178.221		185.940	185.940		
Pb			220.353	220.353		
Sb		206.833	206.833	206.833		
Si	251.611		288.159		251.611	
Sn	140.052	235.484	189.989	189.989		181.110
Ti	334.941					334.903
Tl		276.787				
V	292.402			310.230		311.838
Zn	206.200	213.856	206.200	206.200	213.856	213.856

7.3.4.3 Noble Metals

Since even extremely small errors may result in a gigantic monetary difference [491], maximum precision (0.1 %) is absolutely demanded [492]. Therefore, the classical gravimetric methods were kept for a long time in noble metal analyses since they were unsurpassed in their accuracy and reproducibility. Only the use of a simultaneously measured internal standard [493, 494, 495, 496] combined with bracketing calibration and multiple calibrations and measurements of the samples provide a similar analytical performance [497]. An excellent summary of noble metal analysis methods with many practical notes is given by Lüschow [498]. The choice of the digestion depends on the metal or alloy to be determined. Gold is dissolved in aqua regia [499].

As well as high precision analysis, purity determinations are carried out. To obtain the best detection limits, axial viewing is required [500]. Line selection in the line-rich matrix of some noble metals and their alloys is not always trivial [501, 502]. This is aggravated by the fact that solutions with very high concentrations of matrix are prepared (up to 50 g/L) to obtain the lowest possible limits of detection with respect to the solid sample. In addition, at such high concentrations even less sensitive lines may interfere by spectral overlap. Another great dificulty consists of the fact that the practical limits of detection in the solutions are deteriorated by contamination by ubiquitous elements (such as Ca, Mg, and Si). Therefore, special attention has to be paid to the highest purity during the complete analytical procedure. Some potentially useful analytical lines are listed in Table 18.

The element Os in its highest oxidation state as OsO_4 is both highly volatile and highly toxic! In addition, it tends to give severe memory-effects in the sample introduction system.

7.3.4.4 Special Alloys

This large group of applications is so specialized that dealing with them as a whole does not seem possible in the context of this book. Hence, it does not seem sensible to provide a table with analytical lines in the same way as has been done for many other groups. Numerous applications are documented [503, 504, 505, 506, 507]. In order to be able to measure correctly in the often very line-rich matrices, the application of multivariate regression techniques for the correction of spectral interference is recommended [508, 509, 510, 511]. After the dissolution step, some metals will be present as anions. These can easily be separated from the rest of the matrix to facilitate determination of the metals [512].

Table 18: Analytical lines for noble metals and their alloys

Element	Refined silver wavelength [nm]	Ag-Cu-alloy wavelength [nm]	Refined gold wavelength [nm]	Jewelry gold wavelength [nm]	Dental alloys wavelength [nm]
Ag		328.068	328.068	328.068	328.068
Al	396.152		396.152		
As	188.979		188.979		
Au	242.795	242.795		242.795	242.795
B	249.773		249.773		
Bi	190.171		223.061		
Ca			396.847		
Cd	228.802	226.502	228.802	226.502	
Co	228.616		238.892		
Cr	205.560		357.869		
Cu	327.396	327.396	327.396	327.396	327.396
Fe	259.940		259.940		259.940
Ga	294.364	294.364	294.364	294.364	
In			303.936		325.609
Ir	224.268	205.222	237.277	205.222	205.222
K	769.896		766.490		
Mg	285.213		279.553		
Mn	260.568		257.610		
Mo			281.616		
Na	588.995		589.592		
Ni	221.648	231.604	341.476	227.021	
Pb	220.353	220.353	220.353	283.306	
Pd	340.458	340.458	340.458	340.458	340.458
Pt	214.423	217.467	214.423	217.467	217.467
Rh	343.489	343.489	343.489	343.489	
Ru	349.894	349.894	240.272	349.894	
Sb	217.582		217.582		
Se	196.026		196.026		
Si	288.158		251.611		
Sn	283.998	235.484	189.927	235.484	235.484
Te	214.281		214.281		
Ti	336.121		334.940		
W			207.912		
Zn	206.200	206.200	206.200	206.200	206.200
Zr			343.823		

7.3.5 Material Sciences

7.3.5.1 Semiconductors

As a rule, in the semiconductor industry the absence of harmful elements in a material has to be proven. Even at trace and ultra-trace level, some elements disturb the functionality of the electronic component and their absence must be guaranteed [513, 514]. Many analyses are done by ICP-MS because of its greater power of detection. However, in ICP-OES far higher matrix concentrations in the sample solution can be aspirated. Therefore, the difference between these techniques is sometimes not great for some elements, and ICP-OES may even have a slight advantage. The strongest emission lines are frequently selected for the analysis. The element B can be determined after distilling it off as the fluoride [515].

7.3.5.2 Ceramics

Ceramics are used in household crockery, high temperature ceramics, structural components, and in a fast growing market as a carrier material for catalysts [516]. Some metals carried on the surface of a ceramic catalyze chemical reactions. Other metals inhibit these reactions ("catalyst poison"). Therefore, it is necessary to determine the concentrations of the desired and of the unwanted metals. The material can be digested by lithium tetraborate fusion. In order to obtain a clean solution, a 10-fold surplus of the digestion reagent is necessary [517]. Ceramic catalysts can also be digested by a mixture of hydrofluoric acid, hydrochloric acid, nitric acid and sulfuric acid, which is heated to dryness [518]. With the aid of a microwave-assisted digestion, most ceramics can be dissolved in similar acid mixtures in a closed system without the risk of losses [519].

Table 19: Analytical lines for ceramics used as a catalyst base

Element	Wavelength [nm]
Al	396.152
Ca	317.937
Fe	238.204
K	766.491
Mg	279.553
Mo	202.030
Na	589.541
Pb	220.353
Si	212.412
Ti	334.941
V	292.402
W	207.911

7.3.6 Industrial Applications

7.3.6.1 Industrial Chemicals and Fertilizers

This group covers a wide spectrum. Therefore, only a few examples are listed here.

- Mineral fertilizer is digested with a mixture hydrochloric and nitric acids [520].
- Flotation concentrates for the production of lead and zinc are digested after a pretreatment with hydrochloric acid in a mixture of nitric and perchloric acids. After the mixture has been evaporated nearly to dryness, it is dissolved in a solution of EDTA [521].
- For the purity determination of phosphoric acid (85 %), the sample is diluted 1 + 2 with 18 MOhm water and is measured in "hot" plasma conditions (RF power: 1500 W, nebulizer gas flow: 0.65 L/min).
- Cosmetic products are analyzed directly or after dilution with water [522].
- For the purity determination of chemicals for the production of batteries (e.g. Zn or MnO_2), the metals or their compounds are measured in matrix concentrations of 2 to 5 %.

A review is given in [523].

Table 20: Analytical lines for samples from different production processes

Element	Galv. baths, wavelength [nm]	Cement, wavelength [nm]	Glass, wavelength [nm]
Ag	328.068		396.152
Al	396.152	396.153	
B			249.773
Ba	455.403		455.403
Ca		422.673	393.366
Cd	228.802		228.616
Cr	267.716	205.552	283.563
Cu	324.754		327.396
Fe	259.940	238.204	259.940
K		766.491	766.491
Li			670.781
Mg		279.079	279.553
Mn	257.610	257.610	259.372
Na	589.592	589.592	589.590
Ni	231.604		231.604
P	213.618*	213.618	
Pb	220.353	220.353	220.353
S		180.669	
Si		251.611	251.611
Sn	189.980		
Sr		407.771	407.771
Ti		334.940	334.941
Zn	206.200	213.856	213.856
Zr			343.823

*: Not for baths containing Cu

7.3.6.2 Galvanizing/Electro-plating Baths

The concentrations of both the metals that provide the coating and the inhibiting substances must be present in the bath solution within certain limits [524]. The analysis is somewhat difficult in that the matrix concentration is exceptionally high and its composition is usually unknown. There are some possible solutions, e.g. the use of an internal standard or better a two-stage calibrating method. For the latter procedure, the bath sample is first analyzed as carefully as possible by the method of analyte addition. The concentration determined here is added to the concentrations obtained by the analyte addition. This sum is then used as the new concentration of the perfectly matrix-matched external calibration solution.

Since the compositions of the baths can differ considerably, the wavelengths listed in Table 20 can be used only for the preliminary orientation.

7.3.6.3 Brines and Salts

In these applications, traces are determined to monitor the purity. Therefore, low limits of detection with respect to solid content are desired. One way to achieve this is by preparing solutions with a high salt concentration (20–30 %) [525]. For this, measures to achieve good long-term stability must be taken:

- A nebulizer designed for high dissolved solids is required.
- The injector must be carefully set further away from the plasma than usual to avoid salt deposition at the tip.
- Use a higher RF generator power.
- Install an argon humidifier if necessary.

7.3.6.4 Cement, Gypsum, Calcium Matrix

The cement is digested with a mixture of hydrofluoric acid and aqua regia [526], the surplus hydrofluoric acid being masked with saturated boric acid solution. A common alternative is a lithium metaborate fusion [527] followed by extraction with hydrochloric acid and analysis of the soluble components [528]. Some analytical lines for cement samples are listed in Table 20.

7.3.6.5 Glass

Glass is digested by fusion [529], the sample being mixed with a fivefold excess of lithium metaborate and melted for 15 min at ca. 1000 °C. The melt is poured into 50 mL of 10 % nitric acid with stirring. If the fusion temperature is increased to 1 100 °C and the time is increased to 30 minutes, the excess of lithium metaborate can be reduced to 2 : 1 [530]. Alternatively, samples of glass can be digested with hydrofluoric acid [531]. Table 20 lists some analytical lines for glass samples.

7.3.6.6 Other Industrial Applications

In this category, the industrial applications are summarized which do not fit into any of the previous groups, and the applications are therefore correspondingly various.

An example is the analysis of the toxicologically relevant elements extracted from toys. According to regulation EN 71 Part 3, the amounts of eight elements may not exceed certain limits [532]. The control can be done with ICP-OES [533].

Another example is the analysis of materials from the nuclear industry [534, 535].

Table 21: Analytical lines for eluates of toxicologically relevant elements in toys

Element	Wavelength [nm]
As	189.042
Ba	230.424
Cd	226.502
Cr	205.552
Hg	194.227
Pb	220.353
Sb	217.581
Se	196.026

7.3.7. Organic Solvents

If organic solvents are introduced into the plasma, it turns bright green. Its form changes and it becomes a little smaller [536]. The most frequent application is the analysis for elements present in oils, which could be mineral oils (fresh or used) or edible oils. The viscosity of these oils can vary a great deal. The samples (and calibration solutions) are therefore diluted with a solvent to bring down the viscosity so that the sample solution can be pumped by a peristaltic pump, and the influence of the viscosity on the results is also thereby reduced. In addition, "white oil" may be added, mainly in order to match the calibration solutions to the viscosity of the samples. White oil is a pure hydrocarbon mixture, which is available in different viscosities, and is added to the calibration solutions in such an amount that the sum of stock solution and white oil is constant.

Contradictory demands are made of the solvent (for matrix matching). On one hand a low viscosity of the solvent is beneficial to reduce the viscosity of the mixture, but on the other hand a high boiling-point will lead to a more stable plasma [537]. The lower the boiling point or boiling range range, the more solvent vaporizes and reaches the plasma. Just to recall: the amount delivered into the plasma by a pneumatic nebulizer in combination with a Scott nebulizer chamber is only 1–2 % of the amount delivered to the nebulizer by the pump. This is the reason for the (unstable) equilibrium between the energy supplied by the generator via the induction coil and the energy consumed by the material carried into the plasma. If more material is carried into the plasma, it will be destabilized and in the worst case will be extinguished.

When choosing a solvent, one should prefer aliphatic hydrocarbons, as these are generally less toxic than aromatic compounds. A further important aspect is the fact that unsaturated compounds (and especially aromatic hydrocarbons) tend to forms carbon deposits, which impair the long-term stability of the system. Matrix interference was minimized using tetralin [538].

The main aim when starting an application with a new organic solvent is to ensure that the plasma remains stable. The first and foremost option is to reduce the amount of solvent that ultimately enters the plasma, and the first step to take is to minimize the pump rate so that less solution is transported to the nebulizer. When using highly volatile solvents, an injector with a small interior diameter should be inserted [539]. The shape of the tip of the injector can also have an influence on the deposits formed on it [540].

Furthermore, the nebulizer gas flow should be reduced since the vapor evolved from the solvent contributes to the volume flow. Practically, one proceeds as follows. Turn down the nebulizer gas flow so that the aerosol does not reach the plasma (typically this happens at about 0.2 L/min). Then slowly increase the gas flow in very small steps (approx. 0.02 L/min) and wait a few minutes after each step. If the plasma is extinguished, displace the remaining vapors by purging with nebulizer gas but without pumping or aspirating the solvent any further. Then return to the safe setting determined previoiusly and ignite the plasma again. Raise the nebulizer flow until the initial radiation zone becomes visible between the turns of the induction coil or slightly above it.

The proportion of evaporated solvent can be further reduced by cooling the nebulizer chamber. In the simplest case, a tube attached to a cryostat is wrapped around the nebulizer chamber. In some applications, cooling to −30 °C is required. Some manufacturers of instruments or accessories offer cooled nebulizer chambers (Fig. 127).

In the case of a spectrometer whose nebulizer chamber is insufficiently protected against heat radiation from the plasma, it can in some cases be sufficient to attach the nebulizer chamber outside this torch area. The outlet of the nebulizer chamber is connected by a Tygon tube of a large diameter to the bottom of the injector. Some instruments on the market have a heated nebulizer chamber. Of course, this additional heat source must be turned off.

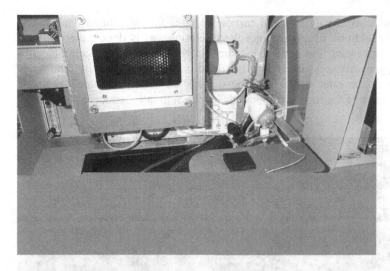

Fig. 127: The photograph shows a cooled nebulizer chamber (center right) for the analysis of highly volatile organic solvents (source: Thermo Elemental)

Contrasting with cooling the nebulizer chamber for the analysis of highly volatile solvents, an application has been described in which the sample supply tube was heated in order to reduce the viscosity of the oils. Thus, the dilution step can be abandoned [541, 542]. Here the complete length of the tube from the sample container to the nebulizer and thence to the nebulizer chamber is heated to a constant temperature.

Speeding up the pump in order to shorten the delay time as is sometimes done with aqueous samples should not be done in the case of organic solvents. With a fast pump speed, the quantity of the solvent introduced suddenly increases and adversely affects the stability of the plasma, which can be extinguished in the worst case.

Especially when using a solvent with a low boiling point, a higher RF power should be set to provide sufficient energy [543, 544]. Adjust the RF power in 100 W steps to higher settings in cases where the plasma is not stable enough. In addition, an improvement in the plasma stability can be achieved with an increased coolant flow.

If the plasma is then stable, carbon deposits may form at the tip of the injector or in rare cases at the intermediate tube of the torch. A longer distance between injector and plasma helps in this case. Even fractions of a millimeter can give a significant improvement. The method of increasing this distance varies from instrument to instrument. One can either do it by mechanically moving the injector or the sample introduction system or by a setting a higher auxiliary gas flow. In a number of instruments, both can be changed. Of course, the viewing height must then be re-optimized (for radial viewing). Carbon deposits can be cleaned off by putting the injector into a muzzle furnace and heating to 550 °C for several hours [545].

Carbon deposits do not appear if the carbon is burnt away completely by oxygen. Therefore short torches (that are only just long enough to shield the induction coil) or extended torches with several slits should generally be preferred. The direct addition of oxygen is another successful way to avoid carbon deposits and structured background (Fig. 128) [546]. The addition of oxygen into the operating gas is done either with an additional regulator unit (mass flow controller or needle valve) or by using commercially available argon/oxygen mixtures. If oxygen is added as an extra gas, the amount is slowly raised while at the same time observing the carbon band. No more oxygen is added once the influence of the carbon compounds has been greatly reduced.

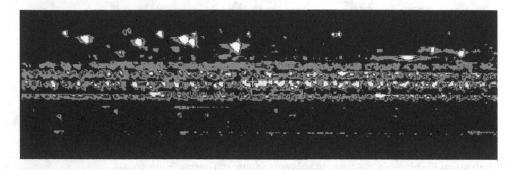

Fig. 128: The two Echellograms of the plasma emission in the spectral range 400–600 nm clearly demonstrate the effect of oxygen addition to the plasma: In the upper photograph, C_2 molecular bands can be seen over a wide spectral range. These have disappeared in the lower photo after the addition of 100 mL/min oxygen. The remaining lines are predominantly argon emission lines (source: Thermo Elemental)

The choice of suitable pump tube material can presents a great challenge from time to time since some solvents are so powerful that they dissolve the tube material in a relatively short time. The recommendations of the manufacturers of pump tubes should be taken into account here. It might be considered to try to avoid a pump altogether for some very potent solvents. However, the viscosity must be matched perfectly in this case.

When working with organic solvents, the importance of the chemical bond of the analytes in standards and samples is generally greater than in aqueous systems. This can have a considerable influence on the accuracy. A spiked sample should be checked to determine whether the added substance can be accurately determined. In addition, wherever possible, calibration standards from different manufacturers should be used. For some applications (e.g. determination of S), the only way to get correct results is to use standard reference materials (e.g. from NIST).

The stability of the samples, the diluted sample solutions, and the prepared standard solutions is clearly lower than that of aqueous solutions.

7.3.7.1 Wear Metals and Contamination in Oil

Used oils often contain particles, which are deposited in the sample container or the nebulizer chamber and are therefore not detected during a direct analysis. Of the particles that reach the plasma, only the very small ones become atomized or ionized completely because of the short dwell time in the plasma [547].

The best results for very viscous used oils have been obtained with large dilution factors in the range 1 : 50 to 1 : 100.

In many cases, an analysis for abraded metal or for elements that contaminate the oil involves looking for a trend. Increasing or decreasing concentrations of certain elements indicate a future malfunction of an engine, for example. Here, accuracy is not as important, and a systematic error would probably be acceptable. However, it is important to ensure that the current measurement has the same error as all the earlier ones [548].

Since carbon fragments such as C_2 have an extensive range of molecular emission bands between 400 and 600 nm and there are also CO bands at 230 nm [549], some analytical wavelengths, which have proven to be useful for aqueous samples, cannot be used. Table 22 gives a list of wavelengths, which have proved useful in practice.

7.3.7.2 Additives

The aim of the analysis of oil additives is to monitor compliance with specifications. Here, good precision is required. Since the concentrations are usually very high, the dilution factor must also be very high (e.g. 1 : 1000 or even more). The advantage of this high dilution is the fact that matrix effects disappear. On the other hand, the dilution error becomes more significant. Table 22 gives a selection of wavelengths which have been found useful.

7.3.7.3 Tar

Tar samples must be very much diluted to reduce the viscosity to enable the solution to be pumped. Digestion is a better approach [550].

Table 22: Analytical lines for wear metals, contaminants and additives in oils

Element	Wavelength [nm]	Alternative wavelength [nm]	Further wavelength [nm]
Ag	328.068		
Al	396.152	308.215	
B	249.773		
Ba	455.403	233.527	493.408
Ca	396.847	422.673	317.933
Cd	228.802		
Cl	134.724	135.165	136.345
Cr	267.716	283.563	
Cu	327.393	324.754	
Fe	259.940		
Mg	279.553	279.079	285.213
Mn	257.610		
Mo	281.615	202.030	
Na	589.592	588.995	
Ni	221.648	341.476	
P	213.617	178.221	214.914
Pb	220.353	283.306	
S	180.734	182.037	
Si	251.611	288.158	
Sn	189.927	283.998	
Ti	334.941	337.279	
V	309.311	311.071	
Zn	206.200	213.856	334.501

7.3.7.4 Edible Oils

Many edible oils can be easily analyzed after diluting with petroleum or kerosene (at least by a factor of 2). However, a resin may form at the tip of the injector in the case of some oils. A long-term stability test is therefore required (minimum about 4 h) in order to check the suitability of ICP-OES for this application.

8 Procurement of Equipment and Preparation of the Laboratory

8.1 Which Atomic Spectrometric Technique is the most Suitable?

When considering the purchase of an ICP emission spectrometer, the very first question should be: "Which analytical technique will best suit the application?" Is ICP-OES really the best choice or could the job be done better with some other atomic spectrometric technique such as AAS or ICP-MS [551, 552, 553]? With analytical instruments as with everything else, nothing can do everything, i.e. no single piece of equipment can perform every conceivable analytical task. We may wish for something that can do everything (Fig. 129) but it does not exist, so although it is true that manufacturers have steadily improved the analytical performance of ICP emission spectrometers, these are often specialized for particular applications.

Fig. 129: What we want...   ...and what we can have.
(Dream and reality as portrayed by Albert Grundler)

To pinpoint the most suitable technique, the immediate analytical task should be described as precisely as possible and an estimate of any future increase in the job requirements should be made to try to avoid acquiring an instrument which will not meet changing demands. These points should be raised:

- How many elements have to be determined?
- Which elements have to be analyzed?
- What is the concentration range?
- What is the matrix and could it interfere?
- What amount of sample is available?

As a next step, the general current analytical requirements, precision (accuracy, reproducibility), analysis speed and throughput, number of the determinations (elements, samples) and costs (acquisition costs, operating costs) must be considered. The specific analytic task will then be evaluated according to these criteria.

The general characteristics of the techniques will at once point in a certain direction. Flame AAS is best for a small number of elements at high concentration to be determined in a simple matrix. Graphite furnace AAS tolerates a high matrix concentration and is suitable for a few elements in the trace range. ICP-OES generally covers the concentration range of both these AAS techniques for the determination of a large number of elements with a medium matrix load. ICP-MS has the lowest limits of detection. Like ICP-OES, the technique is suitable for the determination of many elements. However, the matrix load should be very low. Using the hydride or cold vapor techniques, the limits of detection can be improved for some elements by around an order of magnitude.

If only a small number of elements per sample have to be determined, this points to AAS. ICP technologies have an advantage if the number of elements is about five or more. The question of which elements suit which technique is more difficult to answer. The periodic table in Fig. 130 gives a general indication of this.

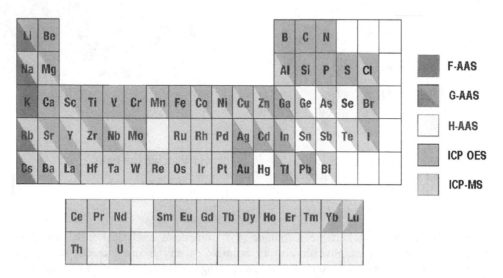

Fig. 130: A general assessment of which element should be analyzed preferentially with which atomic spectrometric technique.

As far as accuracy is concerned, absence of interference is as important as limits of detection. The technique with the best detecting power is ICP-MS; ICP-OES and graphite furnace AAS come below this, and flame AAS is only usable for relatively high concentrations. The linear working range of up to about six orders of magnitude is clearly bigger for the two ICP techniques than for AAS.

Interference by the matrix can have different causes. Chemical interference appears primarily in AAS, while sample transport interference is mainly observed in ICP-OES. Spectral interference is particularly common in the two ICP techniques.

The graphite furnace technique requires the lowest amount of sample (typically 20–100 µL), the ICP technologies require more (a few mL), and flame AAS the most (ca. 10 mL).

If we compare measurement speeds, we find that flame AAS is the fastest. A single determination takes only about 10–20 s. However, this is an element-oriented technique. Therefore, the time for all the results of a multi-element analysis to be ready may be quite considerable. A measurement cycle typically takes a few minutes by all other techniques. However, in the graphite furnace technique only one determination will be complete after this period of, while in the ICP techniques – depending on instrument type – the results of a triple determination of some dozen elements will have been measured by this time. One characteristic whose influence on the analysis speed is commonly underestimated is the working range. ICP techniques have large working ranges, so that a lot of time is saved by not having to dilute the samples.

This has a direct influence on the running costs, and the time for the instrument to amortize the high acquisition cost is often lower than anticipated. The acquisition cost of an ICP mass spectrometer is the highest, followed by that of an ICP emission spectrometer. Atomic absorption spectrometers with a graphite furnace furnace are a little less expensive and flame AAS instruments costs by far the least. The operating costs are in general highest for graphite furnace AAS because of the cost of the graphite tubes, while the cost of the argon is an important factor in the running costs of the ICP techniques. From the point of view of running costs, flame AAS is again the least costly technique.

The skill of the user is mentioned here only to complete the picture. Even though flame AAS is considered the simplest technique; there are a number of sources of interference which can cause erroneous results if the user does not take effective counter-measures. All other techniques are classified as being more demanding. The success or failure then depends on circumstances (type of application and instrument). A thorough training is suggested.

8.2   Which ICP Emission Spectrometer is the Most Suitable?

If the decision has been made in favor of ICP-OES, then the next question is which ICP emission spectrometer is best to perform the analytical task to hand. This evaluation should be carried out by measuring samples, which best represent the intended application. If there are many intended laboratory applications, it is meaningful to select the most important and/or challenging ones. A few carefully thought out and well-prepared tests will yield more information to enable a good evaluation to me made than a multiplicity of random results which are difficult to interpret. With the specific tasks in mind, a test program should be developed which includes the following criteria:

- Accuracy
  - Resolution
  - Non-spectral interference
- Reproducibility
- Sensitivity (BEC)
- Limits of detection
- Long-term stability
- Analysis speed and sample throughput
- User friendliness
  - Hardware
  - Software
- Costs
  - Acquisition
  - Running costs.

The accuracy is best checked with certified standard reference materials. The comparison with data gained with the instrument currently used, may (but need not) agree with the "true" results. If there is no other option, one should analyze the samples for the test very thoroughly. At least one sample should contain a spike of the most important analytes. As a supplement, the resolution should be checked. It must be borne in mind that the resolution is wavelength dependent in some systems (Echelle) , so that line profiles should be taken over the complete spectral range. In addition, it should be checked how the instrument reacts to non-spectral interference. Samples with a higher salt load or higher acid concentration than usual can indicate that there are sources of interference from sample transport or excitation.

The reproducibility as a statistical quantity is itself subject to certain fluctuations. In order to get a meaningful impression, an average value of all the relative standard deviations of those results which are significantly above the limit of detection (at least a factor of 100) should be calculated.

Since low limits of detection can be obtained only with good sensitivity, one of these tests is redundant. If only high concentrations have to be analyzed, the test may be even superfluous. However, if traces have to be measured, the sensitivity or the limits of detection and the viewing direction (axial or radial viewing) must be considered. If there is good reproducibility, only one of the two parameters (background equivalent concentration or limit of detection) needs to be determined. Determination of the limit of detection (background noise) by the blank method but with the matrix is usually the best approach.

Good long-term stability is then the primary requirement if large numbers of samples are to be analyzed. Here, the method of determination of the long-term stability must be defined before the test. Measurement of a calibration solution (or reference material) at intervals of 15 min for about 4 or 8 h gives the best indication. What is meant by stability should be further clarified by stating the tolerance range: a fluctuation width of 10 % in which all measurements must lie. Alternatively, the results are averaged and the RSD of all measurements can then be consulted as a criterion. If only a small number of samples are to be measured, this test of course becomes superfluous.

For large series of samples, however, long-term stability is an important issue. It should be determined under the conditions of the measurement of the samples. It should be stressed at this point that the determination of sample throughput should be made using the same conditions as those ude in the determination of the limits of detection.

The ease of operation of the ICP spectrometer will frequently not be given a great deal of attention. This is unfortunate because it determines the borderline between joy and frustration in the daily work with the instrument. When inspecting the hardware, you should take a close look at the following:

- How simple is the exchange of components?
  - Is the sample introduction system easy to access?
  - Is it demountable? If yes, how difficult is the re-assembly?
  - Can torch and injector be exchanged separately?
- How are the optics adjusted to the analyte channel?
- How easily can accessories (e.g. ultrasonic nebulizer, solid sampling devices) be attached?
- What temperature requirements does the instrument have for good stability?
- How long does the "start-up" of the system take?
- What must the user be aware of in care and maintenance?
- How goodis repair service? (Consult collegues)

Any assessment of the software is predominantly a question of personal preference. However, the following questions are well worth considering:

- Is operation intuitive?
- Do the sequences of steps in method development and routine analysis follow a logical pattern?
- How robust is the software?
- During method development, what support does the user get from the software
  - How long does the method development take for a model application?
  - How much effort does it take to modify an existing method?
- What options for signal processing are available?
- How simple is routine analysis?
  - Can "urgent samples" be measured in amongst the normal routine?
- Is it possible to jump between tasks?

- Is there an easy to use data transfer utility to transfer data to spreadsheet programs or LIMS?
- How are the data (spectra) archived?
  - Can stored spectra be utilized for recalculating an analysis?
- How is the software maintenance (updating) handled?

Of course, this list is far from complete, but should be extended according to the requirements of the particular laboratory concerned. Examples of possible further topics might be the number of possible calibration solutions, quality control measures....

Sometimes, only the acquisition costs of the instrument itself are taken into account when purchasing an ICP-OES instrument. However, one should remember that an infrastructure must be available in the laboratory (see next section), and the cost of setting this up, which could even involve rebuilding parts of the laboratory, will be in same level as the price of the instrument. Depending on the manufacturer, the instrument price may or may not include such items as:

- A guarantee period
- Installation of the instrument
- Initial training on the instrument (a formal course) and application support.

Furthermore, the costs of consumables should be taken into account:

- Gases
  - Argon (consumption during operation and stand-by. Purity?)
  - Nitrogen (Is it necessary? How much is needed? Purity?)
  - Compressed air (Is it necessary?)
- Materials
  - Torches (useful life time for aqueous solutions, costs)
  - Tubes for peristaltic pump
  - RF generator power tube (useful life time, cost of parts and servicing)
  - Other costs specific for the instrument?
- Energy (mainly for the RF generator).

The diagnostic tips mentioned in Chapt. 6, especially the section "The sensitivity changes significantly (drift)" can be used as an aid to the evaluation of an ICP-OES instrument [554].

8.3 Preparation of the Laboratory

First of all, a space must be provided for the new instrument. Here it is not sufficient to know the dimensions of the spectrometer. Enough room must be reserved for the peripherals. The computer which controls the system is the next important item to be positioned. It should be preferably placed near the sample introduction system, so that the operator can have the "most problematical" part of the instrument under supervision. An auto-sampler should be positioned so that the tube connections can be as short as possible. If accessories such as an ultrasonic nebulizer or a solid sampling device are to be attached, room must of course be provided for these also. These accessories are as a rule used only occasionally, but sufficient space should still be available to install them. In any case, even if no accessories are planned, the instrument should be placed so that the doors of the instrument can be opened to enable available ventilators to provide sufficient fresh air. One should also keep in mind that it must be possible to perform simple maintenance operations, such as a cleaning a ventilator filter, without difficulties. Since there is a chance that any technical instrument could break down, one should clarify which parts of the instrument must be accessible for possible repairs. Finally, one should reserve a space where samples to be measured can be placed near to the spectrometer. However, solutions of any kind should not under any circumstances be put *on* the instrument! Many a serious repair could have been avoided if a sample containing aqua regia had not been spilled over the spectrometer!

Gas connections must be available near the instrument, and details of their specifications should be obtained from the manufacturer well ahead of time to allow enough time for installation. The gas valves should be mounted at a place which can be reached comfortably when the instrument is at its planned location. A central gas supply, if not yet available, is very much to be recommended. Procurement of an automatic valve which changes from the empty to the full argon container when a minimum pressure is reached is worth considering at the sme time as the procurement of the ICP instrument. Otherwise, the user must check frequently, and must if necessary change an empty bottle for a full one. Depending on the usage by the ICP instrument or other instruments that consume a sizeable amount of argon, the user should consider whether liquefied gas would be an option. With a high consumption of argon, liquefied gas is less expensive, more pure and less time and labor intensive.

The exhaust should draw powerfully. Not only the waste gases from the plasma but sometimes also the waste heat from the RF generator must be led away. In order to prevent overheating of the power tube if the performance of the exhaust system deteriorates over time, the exhaust should preferably be rather oversized to be on the safe side. Care should also be taken that the exhaust hood or tubes are not airtight directly over the plasma stand, since oscillations of the exhaust system can then be exacerbated, and the pressure waves caused can be directly transferred to the plasma and lead to a poor reproducibility.

The RF generator and the induction coil are typically cooled by a recirculating cooler. In many regions, environmental considerations and water prices make the use of drinking water for cooling prohibitive. If possible, the recirculating cooler should be banished from the laboratory in order to get rid of the waste heat (and the noise). However, it is important that the cooler should still be visible; otherwise, there is the risk that its maintenance will be forgotten, on the "out of sight out of mind" principle.

Electrical supply lines, fuses, and sockets according to the specifications of the manufacturer must be available before the spectrometer is installed. As with the gas supply, enough time should be reserved for their installation.

When planning the ICP-OES instrument, one should also consider the need for other supplies. These include pure water (and even ultra-pure water if this is required for the application) and an adequate quantity of pure sample containers and stock solutions for calibration solutions. One will then be fully prepared for the first attempts to operate the new ICP emission spectrometer as soon as possible after its installation.

After reading this book, nothing really should go wrong!

9 References

[1]   International Organization for Standardization (ISO) (Ed.): "ISO 11885 – Water quality – The determination of 33 elements by inductively coupled plasma atomic emission spectroscopy", 1996.

[2]   USEPA-ICP Users Group (Edited by T. D. Martin et al.): "Method 200.7 – Determination of metals and trace elements in waters and wastes by inductively coupled plasma-atomic emission spectrometry", Cincinnati, Ohio, USA 1994.

[3]   K. Ohls: "30 Jahre hochfrequente Plasmaflammen", Nachr. Chem. Tech. Lab. 41 (1993), 572–578.

[4]   S. Greenfield, I. L. I. Jones, C. T. Berry: "High Pressure Plasmas as Spectroscopic Emission Sources", Analyst 89 (1964), 713–720.

[5]   R. H. Wendt, V. A. Fassel: "Inductively-Coupled Plasma Spectrometric Excitation Source", Anal. Chem. 37 (1965), 920–922.

[6]   V. A. Fassel: "Quantitative Elemental Analyses by Plasma Emission Spectroscopy", Science 202 (1978), 183–191.

[7]   P. W. J. M. Boumans: "Inductively Coupled Plasma-Atomic Emission Spectroscopy: Its Present and Future Position in Analytical Chemistry", Fresenius Z. Anal. Chem. 299 (1979), 337–361.

[8]   S. Greenfield: "Plasma Spectroscopy Comes of Age", Analyst 105 (1980), 1032–1044.

[9]   R. M. Barnes: "Inductively coupled plasma emission spectroscopy: a review", Trends in Analytical Chemistry 1 (1981), 51–56.

[10]   J. Nölte: "ICP-OES: Entwicklungen in den letzten 25 Jahren", CLB Chemie in Labor und Biotechnik 51 (2000), 286–292.

[11]   J. A. C. Broekaert: "Entwicklungstendenzen in der Atomabsorptions- und Plasmaspektrometrie", GIT Fachz. Lab. 38 (1994), 1049–1052.

[12]   P. Rommers, P. Boumans: "ICP-AES versus (LA-)ICP-MS: Competition or a happy marriage? – A view supported by current data", Fresenius J. Anal. Chem. 355 (1996), 763–770.

[13]   Z. Grobenski, J. Nölte: "Recent Innovations in the Inductively-Coupled Plasma Emission Spectroscopy (ICP-OES)", Kem. Ind. 46 (1997), 113–117.

[14]   International Union of Pure and Applied Chemistry – Analytical Chemistry Division: "Nomenclature, Symbols, Units and Their Usage in Spectrochemical Analysis – III. Analytical Flame Spectroscopy and Associated Non-flame Procedures", Pure & Appl. Chem. 45 (1976), 105–123.

[15]   ISO/CD 12235 "General guidelines for inductively coupled plasma emission spectrometry", final draft 1996.

[16]   B. Welz, M. Sperling: "Atomic Absorption Spectrometry", Wiley-VCH, Weinheim 1998.

[17]   G. Schlemmer, B. Radziuk: "Analytical Graphite Furnace Atomic Absorption Spectrometry: A Laboratory Guide", Birkhäuser, Basel 1999.

[18]   J. C. Van Loon: "Atomic Fluorescence Spectrometry – Present Status and Future Prospects", Anal. Chem. 53 (1981), 332A–344A.

[19]   D. R. Demers, Ch. D. Allemand: "Atomic Fluorescence Spectrometry with an Inductively Coupled Plasma as Atomization Cell and Pulsed Hollow Cathode Lamps for Excitation", Anal. Chem. 53 (1981), 1915–1921.

[20]   J. Sneddon: "Direct Current Plasma: A Versatile Excitation Source for Spectrochemical Analysis", Spectrosc. Int. 1 (1989), 24–35.

[21]   F. Leis, J. A. C. Broekaert, H. Waechter: "A three-electrode direct current plasma as compared to an inductively coupled argon plasma", Fresenius Z. Anal. Chem. 333 (1989), 2–5.
[22]   Anonymous: "dc Plasma Emission – An Analytical Technique", Technical Report #307, Spectrametrics, Andover, MA, USA.
[23]   C. I. M. Beenakker, P. W. J. M. Boumans, P. J. Rommers: "Ein durch Mikrowellen induziertes Plasma als Anregungsquelle für die Atomemissionsspektrometrie", GIT Fachz. Lab. 25 (1981), 82–87 und 179–189.
[24]   G. Volland, P. Tschöpel, G. Tölg: "Bestimmung von Elementspuren im ng- und pg-Bereich durch optische Emissionsspektrometrie mit He-MIP-Anregung nach elektrolytischer Abtrennung im Graphitrohr und nachfolgender elektrothermischer Atomisierung", Spectrochim. Acta 36B (1981), 901–917.
[25]   B. D. Quimby, P. C. Uden, R. M. Barnes: "Atmospheric Pressure Helium Microwave Detection System for Gas Chromatography", Anal. Chem., 50 (1978) 2112-2117.
[26]   H. A. Dingjan, H. J. de Jong: "A comparative study of two cavities for generating a microwave induced plasma in helium or argon as a detector in gas chromatography", Spectrochim. Acta 36B (1981), 325–331.
[27]   A. L. Pires Valente, P. C. Uden: "Comparison of a Combined Helium-argon Plasma With Pure Helium Plasmas for Gas Chromatography With Atomic Emission Detection", Analyst 120 (1995), 419–421.
[28]   A. Disam, P. Tschöpel, G. Tölg: "Emissionsspektrometrische Bestimmung von Elementspuren in wässrigen Lösungen mit einem mantelgasstabilisierten, kapazitiv angekoppelten Mikrowellenplasma (CMP)", Fresenius Z. Anal. Chem. 310 (1982), 131–143.
[29]   G. Wünsch, N. Czech, G. Hegenberg: "Bestimmung von Wolfram mit dem kapazitiv gekoppelten Mikrowellenplasma", Fresenius Z. Anal. Chem. 310 (1982), 62–69.
[30]   D. Sommer, J. Flock: "Glimmlampen- und Funkenemissionsspektrometrie – Ersatz oder Ergänzung", GIT Fachz. Lab. 40 (1996), 508–515.
[31]   J. A. C. Broekaert, R. Klockenkämper, J. B. Ko: "Emissionsspektrometrische Präzisionsbestimmung der Hauptbestandteile von Cu/Zn-Legierungen mittels Glimmlampe und ICP", Fresenius Z. Anal. Chem. 316 (1983), 256–260.
[32]   Karl A. Slickers: "Die automatische Emissions-Spektralanalyse", Bühlsche Universitätsdruckerei, Gießen 1977.
[33]   DIN 51008-1 "Optische Atomemissionsspektralanalyse (OES) – Teil 1: Systeme mit Funken und Niederdruckentladungen".
[34]   DIN 51008-1 Beiblatt 1 "Optische Atomemissionsspektralanalyse (OES) – Teil 1: Systeme mit Funken und Niederdruckentladungen".
[35]   Karl A. Slickers: "Spectrochemical analysis in the metallurgical industry", Pure Appl. Chem. 65 (1993), 2443–2452.
[36]   K. Löbe, H. Lucht: "Laserinduzierte Plasmaspektralanalyse zur unmittelbaren Messung fester Proben", GIT Fachz. Lab. 42 (1998), 105–110.
[37]   C. Haisch, U. Panne: "Laser-induced plasma spectroscopy (LIPS) in action", Spectroscopy Eur. 9 {3} (1997), 8–14.
[38]   W. Demtröder: "Laserspektroskopie", Berlin 1993.
[39]   S. Sjöström, P. Mauchien: "Laser atomic spectrocopic techniques – The analytical performance for trace element analysis of solid and liquid samples", Spectochim. Acta Rev. 15 (1993), 153–180.
[40]   B. Német, L. Kozma: "Time-resolved optical emission spectrometry of Q-switched Nd:YAG laser-induced plasmas from copper targets in air at atmospheric pressure", Spectrochim. Acta 50B (1995), 1869–1888.

[41]   L. M. Cabalín, J. J. Laserna: "Experimental determination of laser induced breakdown thresholds of metals under nanosecond Q-switched laser operation", Spectrochim. Acta 53B (1998), 723–730.

[42]   A. I. Whitehouse: "In vessel material analysis of nuclear reactor steam generator tubes", Spectrosc. Eur. 12 {3} (2000), 8–12.

[43]   A. Janssen: "Röntgenfluoreszenzanalyse: Stand der gerätetechnischen Möglichkeiten", LABO (1993), 76–90.

[44]   R. Klockenkämper: "Total-Reflection X-ray Fluorescence Spectrometry: Principles and Applications", Spectrosc. Int. 2 (1987), 26–37.

[45]   K. Flórián, M. Matherny, K. Danzer: "Kritische Bewertung analytischer Methoden anhand metrologischer Charakteristika (Teil 1)", GIT Fachz. Lab. 42 (1998), 693–694.

[46]   F. W. Pinnekamp: "Spektroskopische Messung schneller Plasmaentladungen", GIT Fachz. Lab. 24 (1980), 1026–1030.

[47]   L. Ebdon, S. Greenfield, B. L. Sharp: "The versatile inductively coupled plasma", Chemistry in Britain (Feb. 1986), 123–130.

[48]   M. Sperling: "Optimierung der Anregungsbedingungen in einem induktiv gekoppelten Plasma für die Atomemissionsspektrometrie", Dissertation Universität Hamburg 1986.

[49]   A. Montaser, V. A. Fassel, J. Zalewski: "A Critical Comparison of Ar and Ar-N$_2$ Inductively Coupled Plasmas as Excitation Sources for Atomic Emission Spectrometry", Appl. Spectrosc. 35 (1981), 292–302.

[50]   W. C. Martin, A. Musgrove: "Ground Levels and Ionization Energies for the Neutral Atoms", National Institute of Standards and Technique, Gaithersburg, MD, USA 1999, www@physics.nist.gov.

[51]   R. C. Weast: "CRC-Handbook of Chemistry and Physics", 46th ed., Chemical Rubber Co., Cleveland, Ohio 1964.

[52]   Personal communication by Mr. Petermeier, AirProducts, Hattingen, Germany.

[53]   Anonymous: "Gaslieferservice", LABO (March 2001), 32–33.

[54]   International Union of Pure and Applied Chemistry – Analytical Chemistry Division: "Nomenclature, Symbols, Units and Their Usage in Spectrochemical Analysis – V Radiation Sources", Pure & Appl. Chem. 57 (1985), 1453–1490.

[55]   G. F. Wallace: "Use of a Torch Extension to Reduce ICP Baseline Structure", Atomic Spectrosc. 2 (1981), 93.

[56]   D. J. Devine, R. M. Brown, R. C. Fry: "A Method for Extending or Repairing Inductively Coupled Plasma (ICP) Torches", Appl. Spectrosc. 35 (1981), 332–334.

[57]   R. M. Barnes, R. G. Schleicher: "Temperature and velocity distributions in an inductively coupled plasma", Spectrochim. Acta 36B (1981), 81–101.

[58]   Ch. D. Allemand, R. M. Barnes: "A Study of Inductively Coupled Plasma Torch Configurations", Appl. Spectrosc. 31 (1977), 434–443.

[59]   Ch. D. Allemand, R. M. Barnes, Ch. C. Wohlers: "Experimental studies of reduced size inductively coupled plasma torches", Anal. Chem. 51 (1979), 2392–2394.

[60]   R. N. Savage, G. M. Hieftje: "Characteristics of the background emission spectrum from a miniature inductively-coupled plasma", Analytica Chim. Acta 123 (1981), 319–324.

[61]   L. L. Burton, M. W. Blades: "A Comparison of Excitation Conditions Between Conventional and Low-Flow, Low-Power Inductively Coupled Plasma Torches", Appl. Spectrosc. 40 (1986), 265–270.

[62]   PerkinElmer Corporation: "ICP WinLab32 Software", Norwalk, Conn., USA, 2000.

[63]   P. W. J. M. Boumans: "High Resolution ICP Spectroscopy in Philips Research Laboratories – What has it taught? What has it brought?" ICP Inform. Newsl. 15 (1989), 145–156.

[64]   J.-M. Mermet, C. Trassy: "A spectrometric study of a 40 MHz inductively coupled plasma –V. Discussion of spectral interferences and line intensities", Spectochim. Acta 36B (1981), 269–292.

[65]   T. Matthee, K. Visser: "Background correction in atomic emission spectrometry using repetitive harmonic wavelength scanning and applying Fourier analysis – I. Theory", Spectrochim. Acta 50B (1995), 823–835.

[66]   A. Batal, J.-M. Mermet: "Calculations of some line profiles in ICP-ES assuming Van der Waals potential", Spectrochim. Acta 36B (1981), 993–1003.

[67]   J.-M. Mermet: "Measurement of the Practical Resolving Power of Monochromators in Inductively Coupled Plasma Atomic Emission Spectrometry", J. Anal. At. Spectrom. 2 (1987), 681–686.

[68]   M. W. Blades, G. Horlick: "The vertical spatial characteristics of analyte emission in the inductively coupled plasma", Spectrochim. Acta 36B (1981), 861–880.

[69]   P. W. J. M. Boumans (Ed.): "Inductively Coupled Plasma Emission Spectroscopy, Part 1 Methodology, Instrumentation and Performance", Wiley, New York 1987.

[70]   A. Lopez-Molinero, A. Villareal Caballero, J. R. Castillo: "Classification of emission spectral lines in inductively coupled plasma atomic emission spectroscopy using principal component analysis", Spectrochim. Acta 49B (1994), 677–682.

[71]   J. Jarosz, J.-M. Mermet, J. P. Robin: "A spectrometric study of a 40 MHz inductively coupled plasma-III. Temperatures and electron number density", Spectrochim. Acta 33B (1978), 55–78.

[72]   R. R. Williams, G. N. Coleman: "Ionization of Group II A Elements in the Direct Current Plama: Effects of Ionization Potential on Emission Profiles", Appl. Spectrosc. 35 (1981), 312–317.

[73]   B. L. Caughlin, M. L. Blades: "Analyte ionization in the inductively coupled plasma", Spectrochim. Acta 40B (1985), 1539–1554.

[74]   R. S. Houk: "Model for measuring excitation temperatures on a relative basis without transition probabilities", Spectrochim. Acta 40B (1985), 1517-1524.

[75]   P. A. Abila, C. Trassy: "Rotational Temperatures and LTE in argon ICP", Mikrochim. Acta [Wien] III (1989), 159–168.

[76]   Tetsuya Hasegawa, Hiroki Haraguchi: "A collisional-radiative model including radiation trapping and transport phenomena for diagnostics of an inductively coupled plasma", Spectrochim. Acta 40B (1985), 1505–1515.

[77]   A. Batal, J. Jarosz, J.-M. Mermet: "A spectrometric study of a 40 MHz inductively coupled plasma-VI. argon continuum in the visible region of the spectrum", Spectrochim. Acta 36B (1981), 983–992.

[78]   S. R. Koirtyohann, J. S. Jones, D. A. Yates: "Nomenclature System for the Low-Power argon Inductively Coupled Plasma", Anal. Chem. 52 (1980), 1965–1966.

[79]   S. R. Koirtyohann, J. S. Jones, C. P. Jester, D. A. Yates: "Use of Spatial Emission Profiles and a Nomenclature System as Aid in Interpreting Matrix Effects in the Low-Power argon Inductively Coupled Plasma", Spectrochim. Acta 36B (1981), 49–59.

[80]   G. Horlick, N. Furuta: "Spectrographic observation of the spatial emission structure of the inductively coupled plasma", Spectrochim. Acta 37B (1982), 999–1008.

[81]   D. A. Rodham, J. K. Shurtleff, P. B. Farnsworth: "Energy Transport in the Inductively Coupled Plasma", Mikrochim. Acta [Wien] III (1989), 187–195.

[82]   P. J. Galley, G. M. Hieftje: "The OH 'Bullet' – A Promising Spatial Reference for the Inductively Coupled Plasma", J. Anal. At. Spectrom. 8 (1993), 715–721.

[83]   J.-M. Mermet: "Use of magnesium as a sample element for inductively coupled plasma atomic emission spectrometry diagnostics", Analytica Chim. Acta 250 (1991), 85–94.

[84]   I. B. Brenner, A. T. Zander: "Axially and radially viewed inductively coupled plasmas - a critical review", Spectrochim. Acta, Part B, 55 (2000), 1195–1240.

[85]   L. A. Fernando, N. Kovacic: "Axial Distribution of Analyte Emission in Inductively Coupled argon Plasma", Fresenius Z. Anal. Chem. 322 (1985), 547.

[86]   M. T. C. de Loos-Vollebregt, J. J. Tiggelman, L. de Galan: "End-on observation of a horizontal low-flow inductively coupled plasma", Spectrochim. Acta 43B (1988), 773–781.

[87]   Yoshisuke Nakamura, Katsuyuki Takahashi, Osami Kujirai, Haruno Okochi, C. W. McLeod: "Evaluation of an Axially and Radially Viewed Inductively Coupled Plasma Using an Echelle Spectrometer With Wavelength Modulation and Second-derivative Detection", J. Anal. At. Spectrom. 9 (1994), 751–757.

[88]   F. E. Lichte, S. R. Koirtyohann: "Inductively Coupled Plasma Emission from a Different Angle", ICP Information Newslett. 2 (1976), 192.

[89]   I. B. Brenner, A. Zander, M. Cole, A. Wiseman: "Comparison of Axially and Radially Viewed Inductively Coupled Plasmas for Multi-element Analysis: Effect of Sodium and Calcium", J. Anal. At. Spectrom. 12 (1997), 897–906.

[90]   J. C. Ivaldi, J. F. Tyson: "Performance evaluation of an axially viewed horizontal inductively coupled plasma for optical emission spectrometry", Spectrochim. Acta 50B (1995), 1207–1226.

[91]   D. R. Demers: "Evaluation of the Axially Viewed (End-on) Inductively Coupled argon Plasma Source for Atomic Emission Spectroscopy", Appl. Spectroscopy 33 (1979), 584–591.

[92]   L. J. Prell, C. Monning, R. E. Harris, S. R. Koirtyohann: "The role of electrons in the emission enhancement by easily ionized elements low in the inductively coupled plasma", Spectrochim. Acta 40B (1985), 1401–1410.

[93]   I. B. Brenner, M. Zischka, B. Maichin, G. Knapp: "Ca and Na interference effects in an axially viewed ICP using low and high aerosol loadings", J. Anal. At. Spectrom. 13 (1998), 1257–1264.

[94]   B. Capelle, J.-M. Mermet, J. Robin: "Influence of the Generator Frequency on the Spectral Characteristics of Inductively Coupled Plasma", Appl. Spectrosc. 36 (1982), 102–106.

[95]   M. Sperling: Memo to the Plasmachem-Listserver dated 11 July 2001.

[96]   Ch. D. Allemand, R. M. Barnes: "Design of a Fixed-Frequency Impedance Matching Network and Measurement of Plasma Impedance in an Inductively Coupled Plasma for Atomic Emission Spectroscopy", Spectrochim. Acta 33B (1978), 513–534.

[97]   J. Nölte, Sabine Mann: "Probeneinführungssysteme – die Achillesferse der ICP / Teil 1: Beschreibung der Komponenten", LaborPraxis 25 (March 2001), 42–46.

[98]   R. F. Browner, A. W. Boorn: "Sample Introduction: The Achilles Heel of Atomic Spectroscopy", Anal. Chem. 56 (1984), 786A–798A.

[99]   Sabine Mann, J. Nölte: "Probeneinführungssysteme – die Achillesferse der ICP / Teil 2: Welches Probeneinführungssystem für welche Anwendung?", LaborPraxis 25 (April 2001), 52–56.

[100] G. F. Wallace: "Some factors affecting the performance of an ICP sample introduction system", Atomic Spectrosc. 4 (1983), 188–192.

[101] E. Michaud-Poussel, J.-M. Mermet: "Comparison of nebulizers working below 0.8 L min^{-1} in inductively coupled plasma atomic emission spectrometry", Spectrochim. Acta 41B (1986), 49–61.

[102] M. Tunstall, H. Berndt, D. Sommer, K. Ohls: "Direkte Spurenbestimmung in Aluminium mit der Flammenatomabsorptionsspektrometrie und der ICP-Emissionsspektrometrie – Ein Vergleich", Erzmetall 34 (1981), 588–591.

[103] J. W. Olesik, J. A. Kinzer, B. Harleroad: "Inductively Coupled Plasma Optical Emission Spectrometry Using Nebulizers with Widely Different Sample Consumption Rates", Anal. Chem. 66 (1994), 2022–2030.

[104] I. D. Holclajtner-Antunovic, M. R. Tripkovic: "Study of the Matrix Effect of Easily and Non-easily Ionizable Elements in an Inductively Coupled argon Plasma, Part 2. Equilibrium Plasma Composition", J. Anal. At. Spectrom. 8 (1993), 359–365.

[105] Y. Q. Tang, C. Trassy: "Inductively coupled plasma: the role of water in axial excitation temperatures", Spectrochim. Acta 41B (1986), 143–150.

[106] K. E. LaFreniere, G. W. Rice, V. A. Fassel: "Flow injection analysis with inductively coupled plasma-atomic emission spectroscopy: critical comparison of conventional pneumatic, ultrasonic and direct injection nebulization", Spectrochim. Acta 40B (1985), 1495–1504.

[107] D. R. Wiederin, F. G. Smith, R. S. Houk: "Direct Injection Nebulization for Inductively Coupled Plasma Mass Spectrometry", Anal. Chem. 63 (1991), 219–225.

[108] J. A. McLean, H. Zhang, A. Montaser: "A Direct Injection High-Efficiency Nebulizer for Inductively Coupled Plasma Mass Spectrometry", Anal. Chem. 70 (1998), 1012–1020.

[109] J. S. Becker, H.-J. Dietze, J. A. McLean, A. Montaser: "Ultratrace and Isotope Analysis of Long-Lived Radionuclides by Inductively Coupled Plasma Quadrupole Mass Spectrometry Using a Direct Injection High Efficiency Nebulizer ", Anal. Chem. 71 (1999), 3077–3084.

[110] J. W. Novak, D. E. Lillie, A. W. Boorn, R. F. Browner: "Fixed Crossflow Nebulizer for Use with Inductively Coupled Plasmas and Flames", Anal. Chem. 52 (1980), 576–579.

[111] J. C. Ivaldi, W. Slavin: "Cross-Flow Nebulisers and Sampleing Procedures for Inductively Coupled Plasma Nebulisers", J. Anal. At. Spectrom. 5 (1990), 359–364.

[112] M. A. McLaren, C. Anderau: "The effect of perchloric acid on nebulizer performance", At. Spectrosc. 10 (1989), 77–81.

[113] J. R. Garbarino, H. E. Taylor: "A Babington-type Nebulizer for Use in the Analysis of Natural Water Samples by Inductively Coupled Plasma Spectrometry", Appl. Spectrosc. 34 (1980), 584–590.

[114] P. A. M. Ripson, L. de Galan: "A sample introduction system for an inductively coupled plasma operating on an argon carrier gas flow of 0.1 L/min", Spectrochim. Acta 36B (1981), 71–76.

[115] R. F. Suddendorf, K. W. Boyer: "Nebulizer for Analysis of High Salt Content Samples with Inductively Coupled Plasma Emission Spectrometry", Anal. Chem. 50 (1978), 1769–1771.

[116] M. E. Foulkes, L. Ebdon, S. J. Hill: "Ore and mineral analysis by slurry atomization-plasma emission spectrometry", Anal. Proc. (London) 25 (1988), 92–94.

[117] J. C. Ivaldi, J. Vollmer, W. Slavin: "The Conespray nebulizer for inductively coupled plasma atomic emission spectrometry", Spectrochim. Acta 46B (1991), 1063–1072.

[118] C. S. Saba, W. E. Rhine, K. J. Eisentraut: "Efficiencies of Sample Introduction Systems for the Transport of Metallic Particles in Plasma Emission and Atomic Absorption Spectrometry", Anal. Chem. 53 (1981), 1099–1103.

[119] H. Berndt: "High-pressure nebulization: a new way for atomic spectroscopy", Fresenius J. Anal. Chem. 331 (1988), 321–323.

[120] N. Jakubowski, I. Feldmann, D. Stüver, H. Berndt: "Hydraulic high pressure nebulization – application of a new nebulization system for inductively coupled plasma mass spectrometry", Spectrochim. Acta 47B (1992), 119–129.

[121] W.-D. Drews: "Flüssigkeitszerstäubung durch Ultraschall", Elektronik (1979), 83–86.

[122] R. J. Thomas, C. A. Anderau: "Evaluation of an ultrasonic nebulizer using sequential ICP instrumentation", At. Spectroscopy 10 (1989), 71–73.

[123] I. B. Brenner, P. Bremier, A. Lemarchand: "Performance Characteristics of an Ultrasonic Nebulizer Coupled to a 40.68 MHz Inductively Coupled Plasma Atomic Emission Spectrometer", J. Anal. At. Spectrom. 7 (1992), 819-824.

[124] K. J. Fredeen: "Using ultrasonic nebulization with ICP-AES for analysis of samples with low analyte concentrations", American Laboratory 22 (Dec. 1990), 22-28.

[125] CETAC Technologies Inc.: "Determination of As, Pb, Sb, and Se in Water and Waste-water using ICP-AES and Ultrasonic Nebulization", CETAC Application Note 61, Omaha, USA, 1996.

[126] J. Nölte: „Trinkwasseranalyse per ICP-OES/Ultraschallzerstäubung", LaborPraxis 24 (2000), 30–32.

[127] I. B. Brenner, E. Dorfman: "Application of Ultrasonic Nebulization for the Determination of Rare Earth Elements in Phosphates and Related Sedimentary Rocks Using Inductively Coupled Plasma Atomic Emission Spectrometry with Comments on Dissolution Procedures", J. Anal. At. Spectrom. 8 (1993), 833–838.

[128] W. Schrön, U. Müller: "Influence of heated spray chamber desolvation on the detectability in inductively coupled plasma atomic emission spectrometry", Fresenius J. Anal. Chem. 357 (1997), 22–26.

[129] Anonymous: "Improved Detection Limits for Trace Metals in Organic Solvents by Quadrupole ICP-MS ", CETAC Application Note 114, Omaha, USA.

[130] Anonymous: "Improved Quadrupole ICP-MS Detection Limits for Calcium, Iron and Potassium", CETAC Application Note 134, Omaha, USA.

[131] B. L. Caughlin, M. W. Blades: "Effect of wet and dry nebulizer gas on the spatial distribution of electron density", Spectrochim. Acta 42B (1987), 353–360.

[132] J. F. Alder, R. M. Bombelka, G. F. Kirkbright: "Electronic excitation and ionization temperature measurements in a high frequency inductively coupled argon plasma source and the influence of water vapour on plasma parameters", Spectrochim. Acta 35B (1980), 163–175.

[133] R. I. Botto, J. J. Zhu: "Use of an Ultrasonic Nebulizer with Membrane Desolvation for Analysis of Volatile Solvents by Inductively Coupled Plasma Atomic Emission Spectrometry", J. Anal. At. Spectrom. 9 (1994), 905–912.

[134] Personal communication Dr. Bauer, Riedwerke, Groß-Gerau, Germany, June 1994.

[135] T. D. Martin, C. A. Brockhoff, J. T. Creed: "Trace metal valence state consideration in utilizing an ultrasonic nebulizer for metal determination by inductively coupled plasma atomic emission spectrometry (ICP-AES)", ICP Inform. Newslett. 19 (1994), 730.

[136] K.-H. Bauer, S. Guzik: "Fehlermöglichkeiten bei der Verwendung des Ultraschallzerstäubers im Rahmen der ICP-OES-Messung von Umweltproben", in: B. Welz (Ed.): CANAS95 Colloquium Analytische Atomspektrometrie, Tagungsband, Überlingen 1996, 341–346.

[137] M. W. Routh: "Characterization of ICP nebulizer aerosols using near-forward angle Fraun-hofer diffraction", Spectrochim. Acta 41B (1986), 39–48.

[138] R. J. Thomas: "Comparison of a pumped drain system with a conventional drain on the Perkin Elmer Plasma II ICP emission spectrometer", At. Spectrosc. 10 (1989), 92–95.

[139] C. S. Saba, W. E. Rhine, K. J. Eisentraut: "Efficiencies of Sample Introduction Systems for the Transport of Metallic Particles in Plasma Emission and Atomic Absorption Spectrometry", Anal. Chem. 53 (1981), 1099–1103.

[140] P. Schramel: "Improvements of ICP-Measurements by Using a Watercooled Spraychamber", Fresenius Z. Anal. Chem. 320 (1985), 233.

[141] J. A. Bertagnolli, D. L. Neylan, D. D. Hammargren: "Effect of Surfactants on ICP Analytical Performance", At. Spectrosc. 14 (1993), 4–7.

[142] S. A. Beres, P. H. Brückner, E. R. Denoyer: "Performance Evaluation of a Cyclonic Spray Chamber for ICP-MS ", At. Spectrosc. 15 (1994), 96–99.

[143] J.-L. Todolí, S. Maestre, J. Mora, A. Canals, V. Hernandis: "Comparison of several spray chambers operating at very low liquid flow rates in inductively coupled plasma atomic emission spectrometry", Fresenius J. Anal. Chem. 368 (2000), 773–779.

[144] Z. Zadgorska, H. Nickel, M. Mazurkiewicz, G. Wolff: "Contribution to the Quantitative Analysis of Oxide Layers Formed on High-Temperature Alloys, Using Inductively Coupled Plasma Atomic Emission Spectroscopy – II. Optimization of the ICP Working Parameters and Typical Analysis of Oxide Layers", Fresenius Z. Anal. Chem. 314 (1983), 356–361.

[145] F. La Torre, N. Violante, O. Senofonte, C. D'Arpino, S. Caroli: "A Novel Approach to the Analysis of Ferritin by Inductively Coupled Plasma – Atomic Emission Spectrometry Combined with High Performance Liquid Chromatography", Spectrosc. Intern. 1 {2} (1989), 46–49.

[146] A. Tomiak: "Erfassung und Auswertung zeitabhängiger Signale bei dem Verbundsystem Gelchromatographie-ICP-OES zur Spurenelementspeziation", GIT Fachz. Lab. 36 (1992), 812–813.

[147] Yi Ming Liu, M. L. Fernández Sánchez, E. Blanco González, A. Sanz-Medel: "Vesicle-mediated High-performance Liquid Chromatography Coupled to Hydride Generation Inductively Coupled Plasma Atomic Emission Spectrometry for Speciation of Toxicologically Important Arsenic Species", J. Anal. At. Spectrom. 8 (1993), 815–820.

[148] J. Schöppenthau, J. Nölte, L. Dunemann: "High-performance Liquid Chromatography Coupled With Array Inductively Coupled Plasma Optical Emission Spectrometry for the Separation and Simultaneous Detection of Metal and Non-metal Species in Soy-bean Flour", Analyst 121 (1996), 845–852.

[149] R. Grümping, A. V. Hirner: "HPLC/ICP-OES determination of water-soluble silicone (PDMS) degradation products in leachates", Fresenius J. Anal. Chem. 363 (1999), 347–352.

[150] W. Tirler, W. Blödorn: „Cr^{3+}/Cr^{6+}-Bestimmung – einfach, schnell und hoch empfindlich", LaborPraxis (Okt. 1998), 52–56.

[151] A. Aziz, J. A. C. Broekaert, F. Leis: "The optimization of an ICP injection technique and the application to the direct analysis of small-volume serum samples", Spectrochim. Acta 36B (1981), 251–260.

[152] Xiao-ru Wang, R. M. Barnes: "Flow Injection ICP On-Line Ion Exchange Analysis with Novel Chelating Resin", ICP Inf. Newsl. 12 (1987), 748.

[153] Y. Israel, A. Lásztity, R. M. Barnes: "On-line Dilution, Steady-state Concentrations for Inductively Coupled Plasma Atomic Emission and Mass Spectrometry Achieved by Tandem injection and Merging-stream Flow Injection", Analyst 114 (1989), 1259–1265.

[154] J. F. Tyson: "Flow injection atomic spectrometry", Spectrochim. Acta Rev. 14 (1991), 169–233.

[155] Z. Horváth, A. Lásztity, R. M. Barnes: "Preconcentration and separation techniques for inductively coupled plasma atomic and mass spectrochemical analyses", Spectrochim. Acta Rev. 14 (1991), 45–78.

[156] G. Wünsch, S. Knobloch, J. Luck, W. Blödorn: "Enrichment procedure for water analysis to fulfill the EPA detection limits with inductively coupled plasma atomic emission spectrometry", Spectrochim. Acta 47B (1992), 199–202.

[157] E. Vereda Alonso, A. Garcia de Torres, J. M. Cano Pavón: "Determination of Nickel in Biological Samples by Inductively Coupled Plasma Atomic Emission Spectrometry After Extraction With 1,5-Bis[phenyl-(2-pyridyl)methylene]thiocarbonohydrazide", J. Anal. At. Spectrum. 8 (1993), 843–846.

[158] R. M. Barnes, P. Fodor, K. Inagaki, M. Fodor: "Determination of trace elements in urine using inductively coupled plasma spectroscopy with a poly(dithiocarbamate) chelating resin", Spectrochim. Acta, Part B, 38/1-2 (1983), 245–257

[159] V. Porta, C. Sarzanni, E. Mentasti: "On-Line Preconcentration and ICP Determination for Trace Metal Analysis", Mikrochim. Acta [Wien] III (1989), 247–255.

[160] N. Prakash, G. Csanády, M. R. A. Michaelis, G. Knapp: "Automatic Off/On-Line Preconcentration for ICP-OES: Powerful Instrumentation for Water Analysis", Mikrochim. Acta [Wien] III (1989), 257–265.

[161] P. Schramel, Li-giang Xu, G. Knapp, M. Michaelis: "Multi-elemental analysis in biological samples by on-line preconcentration on 8-hydroxyquinoline-cellulose microcolumn coupled to simultaneous ICP-AES", Fresenius J. Anal. Chem. 345 (1993), 600–606.

[162] Zhixia Zhuang, Xiaoru Wang, Pengyuan Yang, Chenlong Yang, Benli Huang: "On-line Flow Injection Cobalt-Ammonium Pyrrolidin-1-yldithioformate Coprecipitation for Preconcentration of Trace Amounts of Metals in Waters with Simultaneous Determination by Inductively Coupled Plasma Atomic Emission Spectrometry", J. Anal. At. Spectrom. 9 (1994), 779–784.

[163] Liang Yan-zhong, Yin Ning-wan: "On-line Separation and Preconcentration FIA-ICP-AES System for Simultaneous Determination of REEs and Yttrium in Geological Samples", At. Spectrosc. 16 (1995), 243–247.

[164] K. D. Summerhays, P. J. Lamothe, T. L. Fries: "Volatile Species in Inductively Coupled Plasma Atomic Emission Spectroscopy: Implications for Enhanced Sensitivity", Appl. Spectrosc. 37 {1} (1983), 25–28.

[165] Xiao-ru Wang, R. M. Barnes: "A mathematical model for continuous hydride generation with inductively coupled plasma spectrometry - II. pH dependence of hydride forming elements", Spectrochim. Acta 42B (1987), 139-156.

[166] M. Thompson, B. J. Coles: "Enhanced Sensitivity in the Determination of Mercury by ICPAES", Analyst 109 (1984), 529–530.

[167] R. M. Barnes, Xiao-ru Wang: "Microporous Polytetrafluoroethylene Tubing Gas - Liquid Separator for Hydride Generation and Inductively Coupled Plasma Atomic Emission Spectrometry", J. Anal. At. Spectrom. 3 (1988), 1083-1089

[168] J. Nölte: "Kontinuierliche Hydriderzeugung mit ICP-AES-Detektion", in: B. Welz (Ed.): „6. Colloquium Atomspektrometrische Spurenanalytik", Überlingen 1991.

[169] J. Nölte: "Continuous-Flow Hydride Generation Combined With Conventional Nebulization for ICP-AES Determination", At. Spectrosc. 12 (1991), 199–203.

[170] P. Schramel, L.-Q. Xu: "Determination of arsenic, antimony, bismuth, selenium and tin in biological and environmental samples by continuous flow hydride generation without gas-liquid separator", Fresenius Z. Anal. Chem. 340 (1991), 41–47.

[171] J. Carlos de Andrade, M. Izabel, M. S. Bueno: "A continuous flow cold vapour procedure for mercury determination by atomic emission using the reverse flow injection approach", Spectrochim. Acta 49B (1994), 787–795.

[172] T. R. Gilbert: "Apparatus for the Determination of Mercury by Plasma Emission Spectrometry", Anal. Letters 10 (1977), 599–617.

[173] E. Pruszkowska, P. Barrett, R. Ediger, G. Wallace: "Determination of Arsenic and Selenium Using Hydride System Combined With ICP", At. Spectrosc. 4 (1983), 94–98.

[174] E. de Oliveira, J. W. McLaren, S. S. Berman: "Simultaneous Determination of Arsenic, Antimony, and Selenium in Marine Samples by Inductively Coupled Plasma Atomic Emission Spectrometry", Anal. Chem. 55 (1983), 2047–2050.

[175]   Xiao-ru Wang, Ramon M. Barnes: "A mathematical model for continuous hydride generation with inductively coupled plasma spectrometry analysis - 1. Hydride transfer", Spectrochim. Acta 41B (1986), 967–977.

[176]   A. Menéndez Garcia, J. E. Sánchez Uria, A. Sanz-Medel: "Determination of Arsenic in a Organic Phase by Coupling Continuous Flow Extraction – Hydride Generation With Inductively Coupled Plasma Atomic Emission Spectrometry", J. Anal. At. Spectrom. 4 (1989), 581–585.

[177]   D. T. Heitkemper, J. A. Caruso: "Continuous Hydride Generation for Simultaneous Multi-element Detection with Inductively Coupled Plasma Mass Spectrometry", Appl. Spectrosc. 44 (1990), 228–234.

[178]   X. Wang, M. Viczian, A. Lasztity, R. M. Barnes: "Lead Hydride Generation for Isotope Analysis by Inductively Coupled Plasma Mass Spectrometry", J. Anal. At. Spectrom. 3 (1988), 821–827.

[179]   A. Morrow, B. Wiltshire, A. Hursthouse: "An Improved Method for the Simultaneous Determination of Sb, As, Bi, Ge, Se, and Te by Hydride Generation ICP-AES: Application to Environmental Samples", At. Spectrosc. 18 (1997), 23–28.

[180]   R. C. Campos, C. L. Porto da Silveira, R. Lima: "Detection Limits for Mercury Determination by CV-AAS and CV-ICP-AES Using Gold Trap Preconcentration", At. Spectrosc. 18 (1997), 55–59.

[181]   I. Bertenyi, R. M. Barnes: "Analysis of Trimethylgallium with Inductively Coupled Plasma Spectrometry", Anal. Chem. 58 (1986), 1734-1738.

[182]   A. Lopez Molinero, A. Morales, A. Villareal, J. R. Castillo: "Gaseous sample introduction for the determination of silicon by ICP-AES", Fresenius J. Anal. Chem. 358 (1997), 599–603.

[183]   K.-H. Koch, J. Flock, K. Ohls: "Spurenbestimmung durch die ICP-Aufschlämmtechnik", LaborPraxis {6} (1992), 658–661.

[184]   R. Lobinski, W. Van Borm, J. A. C. Broekaert, P. Tschöpel, G. Tölg: "Optimization of slurry nebulization inductively-coupled plasma atomic emission spectrometry for the analysis of ZrO_2-powder", Fresenius J. Anal. Chem. 342 (1992), 563–568.

[185]   L. Halicz, I. B. Brenner: "Nebulization of slurries and suspensions of geological materials for inductively coupled plasma-atomic emission spectrometry", Spectrochim. Acta 42B (1987), 207–217.

[186]   Les Ebdon, M. E. Foulkes, S. J. Hill: "Direct Atomic Spectrometric Analysis by Slurry Atomisation, Part 9. Fundamental Studies of Refractory Samples", J. Anal. At. Spectrom. 5 (1990), 67–73.

[187]   P. S. Goodall, M. E. Foulkes, Les Ebdon: "Slurry nebulization inductively coupled plasma spectrometry - the fundamental parameters discussed", Spectrochim. Acta 48B (1993), 1563–1577.

[188]   C. J. Walker, D. E. Davey, K. E. Turner: "Effect of slurry mineralogy on slurry ICP-AES", Fresenius J. Anal. Chem. 355 (1996), 801–802.

[189]   K. Knight, S. Chenery, S. W. Zochowski, M. Thompson, C. D. Flint: "Time-resolved Signals from Particles Injected into the Inductively Coupled Plasma", J. Anal. At. Spectrom. 11 (1996), 53–56.

[190]   C. K. Manickum, A. A. Verbeek: "Determination of Aluminium, Barium, Magnesium and Manganese in Tea Leaf by Slurry Nebulization Inductively Coupled Plasma Atomic Emission Spectrometry", J. Anal. At. Spectrom. 9 (1994), 227–229.

[191]   J. Goulter: "Analyse von Neben- und Spurenelementen in Siliciumcarbid mittels Slurry-Technik", Spectro Report Applicationsbericht 63, Kleve, Germany.

[192]   G. Crabi, P. Cavalli, M. Achilli, G. Rossi, N. Omenetto: "Use of the HGA-500 graphite furnace as a sampling unit for ICP emission spectrometry", At. Spectrosc. 3 (1982), 81–88.

[193] R. M. Barnes, P. Fodor: "Analysis of urine using inductively-coupled plasma emission spectroscopy with graphite rod electrothermal vaporization", Spectrochim. Acta Part 38B (1983), 1191-1202.

[194] H. Matusiewicz, R. M. Barnes: "Determination of aluminium and silicon in biological materials by inductively coupled plasma atomic emission spectrometry with electrothermal vaporization", Spectrochim. Acta 39B (1984), 891-899.

[195] A. Golloch, M. Haveresch-Koch, W. G. Fischer: "Anwendung einer elektrothermischen Verdampfungseinheit zu Spurenanalytik", GIT Fachz. Lab. 37 (1993), 91–96.

[196] P. Barth, V. Krivan: "Electrothermal Vaporization Inductively Coupled Plasma Atomic Emission Spectrometric Technique Using a Tungsten Coil Furnace and Slurry Sampling", J. Anal. At. Spectrom. 9 (1994), 773–777.

[197] T. Kántor, G. Záray: "Improved Design and Optimization of an Electrothermal Vaporization System for Inductively Coupled Plasma Atomic Emission Spectrometry", Microchem. J. 51 (1995), 266–277.

[198] Peng Tianyou, Jiang Zucheng: "Fluorination assisted slurry electrothermal vaporization in ICP-AES for the direct analysis of silicon dioxide powder", Fresenius J. Anal. Chem. 360 (1998), 43–46.

[199] D. Marquardt, J. Hassler, P. Perzl: "Spurenanalyse von Feststoffproben: ICP-Spektrometrie mit elektrothermischer Verdampfung", GIT Fachz. Lab. 44 (2000), 206–210.

[200] M. de Laffolie, K. Slickers: "Direkte Analyse von großen und kleinen Metallproben ohne Lösungsvorgang mit Spectroflame-ICP und LISA = Little Samples Analyzer", Spectro Appliaktionsbericht 57/3, Kleve, Germany.

[201] A. Felske, W.-D. Hagenah, K. Laqua: "Über einige Erfahrungen bei der Makrospektralanalyse mit Laserlichtquellen. I. Durchschnittsanalyse metallischer Proben", Spectrochim. Acta 27B (1972), 1–21.

[202] I. Steffan, G. Vujicic: "ICP-AES analysis of nonconductive materials after spark ablation", Spectrochim. Acta 47B (1992), 61–70.

[203] I. B. Brenner, A. Zander, S. Kim, C. Holloway: "Multielement analysis of geological and related non-conducting materials using spark ablation and a sequential spectrometer", Spectrochim. Acta, Part B, 50/4-7 (1995), 565–582.

[204] I. B. Brenner, A. Zander, S. Kim, A. Henderson: "Direct solids analysis of geological and related non-conducting materials using spark ablation and slurry nebulisation", Spectrosc. Eur. 7 {4} (1995), 24–31.

[205] V. Kanicky, J.-M. Mermet: "Selection of internal standards for analysis of silicate rocks and limestones by laser ablation inductively coupled plasma atomic emission spectrometry (LA-ICP-AES)", Fresenius J. Anal. Chem. 355 (1996), 887–888.

[206] P. Goodall, S. G Johnson, E. Wood: "Laser ablation inductively coupled plasma atomic emission spectrometry of a uranium-zirconium alloy: ablation properties and analytical behavior", Spectrochim. Acta 50B (1995), 1823–1835.

[207] R. E. Russo, X. L. Mao, W. T. Chan, M. F. Bryant, W. F. Kinard: "Laser Ablation Sampling with Inductively Coupled Plasma Atomic Emission Spectrometry for the Analysis of Prototypical Glasses", J. Anal. At. Spectrom. 10 (1995), 295–301.

[208] V. Kanicky, V. Otruba, J.-M. Mermet: "Characterization of acoustic signals produced by ultraviolet laser ablation inductively coupled plasma atomic emission spectrometry", Fresenius J. Anal. Chem. 363 (1999), 339–346.

[209] J. Nölte, L. Moenke-Blankenburg, T. Schumann: "Laser solid sampling for a solid-state-detector ICP emission spectrometer", Fresenius J. Anal. Chem. 349 (1994), 131–135.

[210] L. Moenke-Blankenburg, T. Schumann, J. Nölte: "Direct Solid Soil Analysis by Laser Ablation Inductively Coupled Plasma Atomic Emission Spectrometry", J. Anal. At. Spectrom. 9 (1994), 1059–1062.

[211] X. L. Mao, O. V. Borisov, R. E. Russo: "Enhancements in laser ablation inductively coupled plasma-atomic emission spectrometry based on laser properties and ambient environment", Spectrochim. Acta 53B (1998), 731–739.

[212] J. Nölte: "Improving precision for laser solid sampling for ICP emission with an array spectrometer", Fresenius J. Anal. Chem. 355 (1996), 889–891.

[213] J. Nölte, M. Paul: "ICP-OES Analysis Coins Using Laser Ablation", At. Spectrosc. 20 (1999), 212–216.

[214] K. Ohls: "Die Laser-Induzierte-argon-Anregung", LaborPraxis (March 1998), 44–49.

[215] M. Hemmerlin, D. Somas, C. Dubuisson, F. Loisy, E. Poussel, J.-M. Mermet: "Bulk analysis by IR laser ablation inductively coupled plasma atomic emission spectrometry", Fresenius J. Anal. Chem. 368 (2000), 31–36.

[216] J. Nölte: "Spektrale Störungen in der ICP-Emissionsspektrometrie und deren Beseitigung", in: K. Dittrich, B. Welz (Ed.): „CANAS '93", Leipzig 1993, 291–299.

[217] P. W. J. M. Boumans, J. J. A. M. Vrakking: "The widths and shapes of about 350 prominent lines of 65 elements emitted by an inductively coupled plasma", Spectrochim. Acta 41B (1986), 1235–1275.

[218] Perkin Elmer Corporation: "Manual Plasma 400", Norwalk, USA, 1989.

[219] Z. A. Grosser, J. B. Collins: "Determination of the Wavelength Positioning Accuracy of a Sequential Scanning ICP Spectrometer", Appl. Spectrosc. 45 (1991), 993–998.

[220] J. B. Collin, J. C. Ivaldi, M. L. Salit, W. Slavin: "A Spectral Smoothing Filter for ICP-AES", At. Spectrosc. 11 (1990), 109–111.

[221] T. W. Barnard, M. I. Crockett, J. C. Ivaldi, P. L. Lundberg: "Design and Evaluation of an Echelle Grating Optical System for ICP-OES", Anal. Chem. 65 (1993), 1225–1230.

[222] P. N. Keliher, C. C. Wohlers: "Echelle Grating Spectrometers in Analytical Spectrometry", Anal. Chem. 48 (1976), 333A–340A.

[223] G. R. Harrison: "The Production of Diffraction Gratings: II. The Design of Echelle Gratings and Spectrographs", J. Optic. Soc. Am. 39 (1949), 522–528.

[224] Heinz Falk (inventor), Spectro Analytical Instruments, U. S. Patent 5731872 (1998).

[225] M. Duffy, R. Thomas: "Benefits of a Dual-View ICP-OES for the Determination of Boron, Phosphorus, and Sulfur in Low Alloy Steels", At. Spectrosc. 17 (1996), 128–132.

[226] K. W. Busch, M. A. Busch: "Multielement Detection System for Spectrochemical Analysis", Wiley-Interscience, New York 1990.

[227] J. V. Sweedler, R. B. Bilhorn, P. M. Epperson, G. R. Sims, M. B. Denton: "High-Performance Charge Transfer Device Detectors", Anal. Chem. 60 (1988), 282A–291A.

[228] P. M. Epperson, J. V. Sweedler, R. B. Bilhorn, G. R. Sims, M. B. Denton: "Applications of Charge Transfer Devices in Spectroscopy", Anal. Chem. 60 (1988), 327A–335A.

[229] S. W. McGeorge: "Imaging Systems: Detectors of the Past, Present, and Future", Spectrosc. Int. 2 {4} (1987), 26–32.

[230] J. D. Ingle, S. R. Crouch: "Signal-to-Noise Ratio Characteristics of photomultipliers and Photodiodes", Anal. Chem. 44 (1972), 1709.

[231] S. W. McGeorge, E. D. Salin: "Detection Systems for Multielement Analysis with Inductively Coupled Plasma Atomic Emission Spectroscopy", Can. J. Spectrosc. 27 (1982), 25–36.

[232] J.-M. Mermet, J. C. Ivaldi: "Real-time Internal Standardization for Inductively Coupled Plasma Atomic Emission Spectrometry Using a Custom Segmented-array Charge Coupled Device Detector", J. Anal. At. Spectrom. 8 (1993), 795–801.

[233] J. Nölte: "Maßnahmen zur Verbesserung von Richtigkeit und Wiederholbarkeit mit einem Array-ICP-Emissionsspektrometer", in: B. Welz (Ed.): „CANAS '95", Überlingen 1995, 211-220.

[234] J. Nölte: "Spektrometer mit CCD-Detektoren – Neue Möglichkeiten in der ICP-Emission", LaborPraxis 17 (1993), 70–77.

[235] D. Falkin, M. Vosloo: "Charge coupled devices for spectroscopy", Spectrosc. Eur. 5 (1993), 16–22.

[236] V. Karanassios, G. Horlick: "Spectral Characteristics of a New Spectrometer Design for Atomic Emission Spectroscopy", Appl. Spectrosc. 40 (1986), 813–821.

[237] G. M. Levy, A. Quaglia, R. E. Lazure, S. W. McGeorge: "A photodiode array based spectrometer system for inductively coupled plasma-atomic emission spectrometry", Spectrochim. Acta 42B (1987), 341–351.

[238] K. R. Brushwyler, N. Furuta, G. M. Hieftje: "Characterization of a spectrally segmented photodiode-array spectrometer for inductively coupled plasma atomic emission spectrometry", Spectrochim. Acta 46 (1991), 85–98.

[239] J. D. Kolczynski, D. A. Radspinner, R. S. Pomeroy, M. E. Baker. J. A. Norris, M. Bonner Denton, R. W. Foster, R. G. Schleicher, P. M. Moran, M. J. Pilon: "Atomic emission spectrometry using charge injection device (CID) detection", American Laboratory (May 1991), 48–55.

[240] M. J. Pilon, M. B. Denton, R. G. Schleicher, P. M. Moran, S. B. Smith: "Evaluation of a New Array Detector Atomic Emission Spectrometer for Inductively Coupled Plasma Atomic Emission Spectroscopy", Appl. Spectrosc. 44 (1990), 1613–1620.

[241] R. B. Bilhorn, M. B. Denton: "Elemental Analysis with a Plasma Emission Echelle Spectrometer Employing a Charge Injection Device (CID) Detector", Appl. Spectrosc. 43 (1989), 1–11.

[242] R. B. Bilhorn, M. B. Denton: "Wide Dynamic Range Detection with a Charge Injection Device (CID) for Quantitative Plasma Emission Spectroscopy", Appl. Spectrosc. 44 (1990), 1538–1546.

[243] Th. W. Barnard, M. I. Crockett, J. C. Ivaldi, P. L. Lundberg, D. A. Yates, P. A. Levine, D. J. Sauer: "Solid-state Detector for ICP-OES", Anal. Chem. 65 (1993), 1231–1239.

[244] J. Nölte, J. Schöppenthau, L. Dunemann, T. Schumann, L. Moenke-Blankenburg: "Coupling Techniques for Inductively Coupled Plasma Emission Spectrometry Using an Array Spectrometer for Laser Solid Sampling and Speciation", J. Anal. At. Spectrom. 10 (1995), 655–659.

[245] D. Noble: "ICP-AES From Fixed to Flexible", Anal. Chem. 66 (1994), 105A–109A.

[246] P. Boumans: "Developments and Trends in Plasma Spectrochemistry – A View", J. Anal. At. Spectrom. 8 (1993), 767–780.

[247] G. L. Walden, J. N. Bower, S. Nikdel, D. L. Bolton, J. D. Winefordner: "Noise power spectra in the inductively coupled plasma", Spectrochim. Acta 35B (1980), 535–546.

[248] J. Nölte: "Zuverlässige ICP-AES-Meßergebnisse: Optimierung mit Hilfe der gezielten Suchroutine", LaborPraxis 16 (1992), 332–339.

[249] T. Davies, T. Fearn: "Quality control of reference analytical data: How not to fool yourself", Spectrosc. Eur. 8 (1996), 36–39.

[250] P. W. J. M. Boumans: "Measuring detection limits in inductively coupled plasma emission spectrometry using the "SBR-RSDB approach" – I. A tutorial discussion of the theory", Spectrochim. Acta 46B (1991), 431–445.

[251] P. W. J. M. Boumans, J. C. Ivaldi, W. Slavin: "Measuring detection limits in inductively coupled plasma emission spectrometry – II. Experimental data and their interpretation", Spectrochim. Acta 46B (1991), 641–665.

[252]   P. W. J. M. Boumans: "Atomic emission detection limits: more than incidental analytical figures of merit! A tutorial discussion of the differences and links between two complementary approaches", Spectrochim. Acta 46B (1991), 917–939.

[253]   J. Nölte: "Übersichtsanalyse mit dem Plasma 40 unter Nutzung der ICP-Mehrelement-Standardlösungen I bis III" in: B. Welz (Ed.): „5. Colloquium Atomspektrometrische Spurenanalytik", Überlingen 1989, 83–90.

[254]   R. A. Conte, E. H. van Veen, M. T. C. de Loos-Vollebregt: "Fast survey analysis of biomass by-product samples based on ICP optical emission spectra", Fresenius J. Anal. Chem. 364 (1999), 666–672.

[255]   J. A. Morales, E. H. van Veen, M. T. C. de Loos-Vollebregt: "Practical implementation of survey analysis in inductively coupled plasma optical emission spectrometry", Spectrochim. Acta 53B (1998), 683–697.

[256]   R. Degner: "Einfache Bestimmung spektraler Störeinflüsse bei der ICP-Sequenzanalyse am Beispiel der Matrixelemente Fe, Cr, Cu", Fresenius Z. Anal. Chem. 311 (1982), 94–97.

[257]   K. Danzer: "Probenahme inhomogener Materialien – Statistische Modelle und ihre praktische Relevanz", GIT Fachz. Lab. 39 (1995), 928–938.

[258]   D. Truckenbrodt, J. Einax: "Sampling representativity and homogeneity of river sediments", Fresenius J. Anal. Chem. 352 (1995), 437–443.

[259]   S. Mann, J. Nölte: "Impact of high resolution and chemometric interference correction techniques on line selection for ICP emission spectrometry", presented as Poster at the Plasma Winter Conference, Pau, France 1999.

[260]   G. F. Wallace, P. Barret: "Analytical Method Development for ICP Emission Spectrometry", Perkin Elmer Corporation, Norwalk, Conn., USA, 1981.

[261]   K. Danzer, K. Venth: "Multisignal calibration in spark- and ICP-OES", Fresenius J. Anal. Chem. 350 (1994), 339–343.

[262]   A. R. Forster, T. A. Anderson, M. L. Parsons: "ICP Spectra: I. Background Emission", Appl. Spectrosc. 36 (1982), 499–503.

[263]   P. W. J. M. Boumans, J. J. A. M. Vrakking: "Spectral interferences in inductively coupled plasma atomic emission spectrometry – III. An assessment of OH band interferences using the ratio of the limit of determination and the limit of detection as a rational criterion", Spectrochim. Acta 40B (1985), 1423–1435.

[264]   D. Erber, J. Bettmer, K. Cammann: "Einfluß der Blindwertkorrektur auf die Richtigkeit von Messergebnissen in der Routineanalytik", GIT Fachz. Lab. 39 (1995), 340–349.

[265]   R. K. Winge, V. A. Fassel, V. J. Peterson, M. A. Floyd: "ICP Emission Spectrometry: On the Selection of Analytical Lines, Line Coincidence Tables, and Wavelength Tables", Appl. Spectrosc. 36 (1982), 210–221.

[266]   G. R. Harrison: "Massachusetts Institute of Technique Wavelength Tables with Intensities in Arc, Spark, or Discharge Tubes", M. I. I. Press, Cambridge, Massachusetts 1969.

[267]   W. F. Meggers, C. H. Corliss, B. F. Scribner: "Tables of Spectral-Line Intensities – Part I– Arranged by Elements", National Bureau of Standards, Washington, 1961.

[268]   M. L. Parsons, A. Foster, D. Anderson: "An Atlas of Spectral Interferences in ICP Spectroscopy", New York 1980.

[269]   P. W. J. M. Boumans: "Conversion of "Tables of Spectral-Line Intensities" for NBS copper arc into tables for the inductively coupled argon plasma", Spectrochim. Acta 36B (1981), 169–203.

[270]   P. W. J. M. Boumans: "Line Coincidence Tables for Inductively Coupled Plasma Atomic Emission Spectrometry", Pergamon Press, Oxford 1980.

[271]   R. K. Winge, V. J. Peterson, V. A. Fassel: "Inductively Coupled Plasma-Atomic Emission Spectroscopy: Prominent Lines", Appl. Spectrosc. 33 (1979), 206–219.

[272] R. K. Winge, V. A. Fassel, V. J. Peterson, M. A. Floyd: "Inductively Coupled Plasma-Atomic Emission Spectroscopy: An Atlas of Spectral Information", Elsevier, Amsterdam 1985.

[273] C. C. Wohlers: "Experimentally Obtained Wavelength Tables for the ICP", ICP Information Newslett. 10 (1985), 593–688.

[274] P. W. J. M. Boumans, H. Zhi Zhuang, J. J. A. M. Vrakking, J. A. Tielroy, F. J. M. J. Maessen: "Mutual spectral interferences of rare earth elements in inductively coupled plasma atomic emission spectrometry – III. Pseudo physically resolved spectral data: complete results and evaluation", Spectrochim. Acta 44B (1989), 31–93.

[275] G. Wünsch, W. Blödorn, H. M. Ortner: "ICP Wavelength Tables of Tungsten", ICP Information Newslett. 13 (1987), 249–287.

[276] C. Schierle, A. P. Thorne: "Inductively coupled plasma Fourier transform spectrometry: a study of element spectra and a table of inductively coupled plasma lines", Spectrochim. Acta 50B (1995), 27–50.

[277] Z. A. Grosser, J. B. Collins: "Determination of the Wavelength Positioning Accuracy of a Sequential Scanning ICP Spectrometer", Appl. Spectrosc. 45 (1991), 993–998.

[278] J. L. M. de Boer, J. van Leeuwen, U. Kohlmeyer, P. M. Breugem: "The determination of chromium, copper and nickel in groundwater using axial plasma inductively coupled plasma atomic emission spectrometry and proportional correction matrix effect reduction", Fresenius J. Anal. Chem. 360 (1998), 213–218.

[279] D. H. Tracy, S. A. Myers: "Absolute spectral radiance of 27 MHz inductively coupled argon plasma background emission", Spectrochim. Acta 37B (1982), 1055–1068.

[280] G. F. Larson, V. A. Fassel: "Line Broadening and Radiative Recombination Background Interferences in Inductively Coupled Plasma-Atomic Emission Spectroscopy", Appl. Spectrosc. 33 (1979), 592–599.

[281] R. D. Ediger, D. W. Hoult: "Highly structured ICP background emission", At. Spectroc. 1 (1980), 41–47.

[282] J. B. Dawson, R. D. Snook, W. J. Price: "Background and Background Correction in Analytical Atomic Spectrometry – Part 1. Emission Spectrometry – A Tutorial Review", J. Anal. At. Spectrom. 8 (1993), 517–537.

[283] J. Nölte: "Auswahlkriterien für das Setzen von Untergrundkorrekturmeßpunkten in der sequentiellen ICP-AES", Angewandte Atomspektrometrie Nr. 1.5, Überlingen, Germany 1989.

[284] J. Nölte: "Schnell zum sicheren ICP-AES-Meßergebnis mit der automatischen Untergrundkorrektur", CLB Chemie in Labor und Biotechnik 42 (1991), 72–76.

[285] R. I. Botto: "Interference correction for simultaneous multielement determinations by inductively coupled plasma" in: R. M. Barnes (Ed.): "Development in Atomic Plasma Spectrochemical Analysis", Heyden & Sons, Philadelphia 1981, 141–166.

[286] R. I. Botto: "Long Term Stability of Spectral Interference Calibrations for Inductively Coupled Plasma Atomic Emission Spectrometry", Anal. Chem. 54 (1982), 1654–1659.

[287] E. H. van Veen, M. T. C. de Loos-Vollebregt: "Kalman filtering of data from overlapping lines in inductively coupled plasma-atomic emission spectrometry", Spectrochim. Acta 45B (1990), 313–328.

[288] E. H. van Veen, F. J. Oukes, M. T. C. de Loos-Vollebregt: "Some spectral interference studies using Kalman filtering in inductively coupled plasma-atomic emission spectrometry", Spectrochim. Acta 45B (1990), 1109–1120.

[289] E. H. van Veen, M. T. C. de Loos-Vollebregt: "Application of mathematical procedures to background correction and multivariate analysis in inductively coupled plasma-atomic emission spectrometry", Spectrochim. Acta 53B (1998), 639–669.

[290]   E. H. van Veen, S. Bosch, M. T. C. de Loos-Vollebregt: "Kalman filtering approach to inductively coupled plasma atomic emission spectrometry", Spectrochim. Acta 49B (1994), E829–E846.

[291]   J. C. Ivaldi, D. Tracy, Th. Barnard, W. Slavin: "Multivariate methods for interpretation of emission spectra from the inductively coupled plasma", Spectrochim. Acta 47B (1992), 1361–1371.

[292]   J. C. Ivaldi, Th. Barnard: "Advantage of coupling multivariate data reduction techniques with simultaneous inductively coupled plasma emission spectra", Spectrochim. Acta 48B (1993), 1265–1273.

[293]   T. Hoß, S.-A. Ahmadi: "Einsatz simultaner ICP-OES-Messtechnik in der Umweltanalytik", GIT Fachz. Lab. 45 (2001), 732–735.

[294]   K. Danzer, K. Venth: "Multisignal calibration in spark- and ICP-OES", Fresenius J. Anal. Chem. 350 (1994), 339–343.

[295]   J. Nölte: "Minimizing Spectral Interferences with an Array ICP Emission Spectrometer by Using Different Strategies for Signal Evaluation", At. Spectrosc. 19 (1999), 103–107.

[296]   M. R. Tripkovic, I. D. Holclajtner-Antunovic: "Study of the Matrix Effect of Easily and Non-easily Ionizable Elements in an Inductively Coupled argon Plasma. Part 1. Spectroscopic Diagnostics", J. Anal. At. Spectrom. 8 (1993), 349–357.

[297]   B. Budic, V. Hudnik: "Matrix Effects of Potassium Chloride and Phosphoric Acid in argon Inductively Coupled Plasma Atomic Emission Spectrometry", J. Anal. At. Spectrom. 9 (1994), 53–57.

[298]   M. Thompson, M. H. Ramsey, B. J. Coles: "Interactive Matrix Matching: A New Method of Correcting Interference Effects in Inductively Coupled Plasma Spectrometry", Analyst 107 (1982), 1286–1288.

[299]   P. Schramel, J. Ovcar-Pavlu: "Abhängigkeit des Meßsignals von der Säurekonzentration der Probe bei der ICP-Emissionsspektralanalyse", Fresenius Z. Anal. Chem. 298 (1979), 28–31.

[300]   M. Carré, K. Lebas, M. Marichy, M. Mermet, E. Poussel, J.-M. Mermet: "Influence of the sample introduction system on acid effects in inductively coupled plasma atomic emission spectrometry", Spectrochim. Acta 50B (1995), 271–283.

[301]   I.B. Brenner, J.M. Mermet, I. Segal, G.L. Long: "Effect of nitric and hydrochloric acids on rare earth element (REE) intensities in inductively coupled plasma emission spectrometry", Spectrochim. Acta 50B (1995), 323–331.

[302]   J.-L. Todolí, J.-M. Mermet: "Acid interference in atomic spectrometry: analyte signal effects and subsequent reduction", Spectrochim. Acta 54B (1999), 895–929.

[303]   I. I. Stewart, J. W. Olesik: "Transient acid effects in inductively coupled plasma optical emission spectrometry and inductively coupled plasma mass spectrometry", J. Anal. At. Spectrom. 13 (1998), 843–854.

[304]   W. B. Barnett, V. A. Fassel, R. N. Kniseley: "Theoretical principles of internal standardization in analytical emission", Spectrochim. Acta 23B (1968), 643.

[305]   J. C. Ivaldi, J. F. Tyson: "Real-time internal standardization with an axially-viewed inductively coupled plasma for optical emission spectrometry", Spectrochim. Acta 51B (1996), 1443–1450.

[306]   R. M. Belchamber, G. Horlick: "Correlation study of internal standardization in inductively coupled plasma atomic emission spectrometry", Spectrochim. Acta 37B (1982), 1037–1046.

[307]   A. Ryan: "Direct analysis of milk powder on the Liberty Series II ICP-AES with axially-viewed plasma", ICP-AES Instruments at work, Varian Application ICP-21, August 1997.

[308] E. H. Piepmeier, K. Peterson: "Fitting Nonlinear Working Curves for Multiple Analytes When the Amount or Rate of Introduction of Each Standard Differs", Anal. Chem. 66 (1994), 223–229.

[309] J. L. M. de Boer, M. Velterop: "Empirical procedure for the reduction of mixed-matrix effects in inductively coupled plasma atomic emission spectrometry using an internal standard and proportional correction", Fresenius J. Anal. Chem. 356 (1996), 362–370.

[310] M. H. Ramsey. M. Thompson: "Correlated Variance in Simultaneous Inductively Coupled Plasma Atomic-emission Spectrometry: Its Causes and Correction by a Parameter-related Internal Standard Method", Analyst 110 (1985), 519–530.

[311] J. L. M. de Boer, U. Kohlmeyer, P. M. Breugem, T. van der Velde-Koerts: "Determination of total dissolved phosphorus in water samples by axial inductively coupled plasma atomic emission spectrometry", Fresenius J. Anal. Chem. 360 (1998), 132–136.

[312] G. J. Schmidt, W. Slavin: "Inductively Coupled Plasma Emission Spectrometry with Internal Standardization and Subtraction of Plasma Background Fluctuations", Anal. Chem. 54 (1982), 2491–2495.

[313] S. A. Myers, D. H. Tracy: "Improved performance using internal standardization in inductively-coupled plasma emission spectroscopy", Spectrochim. Acta 38B (1983), 1227–1253.

[314] J. Marshall, G. Rodgers, W. C. Campbell: "Myers-Tracy Signal Compensation in Inductively Coupled Plasma Atomic Emission Spectrometry With High Dissolved Solids Solutions", J. Anal. At. Spectrom. 3 (1988), 241–244.

[315] G. F. Wallace: "Some considerations on the selection of an internal standard for ICP emission spectrometry", At. Spectrosc. 5 (1984), 5–9.

[316] M. R. Cave, R. M. Barnes, P. Denzer: "Evaluation and improvement of simplex methods for non-linear optimization and their application to automated ICPES", ICP Inf. Newsl. 7 (1982) 514–515.

[317] L. R. Parker, M.R. Cave, R. M. Barnes: "Comparison of Simplex Algorithms", Anal. Chim. Acta 175 (1985), 231–237.

[318] R. J. Thomas, J. B. Collins: "The Benefits of a Multiparameter Optimization Algorithm for the Analysis of Difficult Samples Using Inductively Coupled Plasma-Atomic Emission Spectrometry", Spectrosc. Int. 5 {2} (1990), 38–45.

[319] S. L. Morgan, S. N. Deming: "Simplex optimization of analytical chemical methods", Anal. Chem. 46 (1974), 1170–1181.

[320] K. Doerffel: "Assuring trueness of analytical results", Fresenius J. Anal. Chem. 348 (1994), 183–187.

[321] B. Griepink, H. Muntau, E. Colinet: "Certification of the Contents of Some Heavy Metals (Cd, Co, Cu, Mn, Hg, Ni, Pb and Zn) in Three Types of Sewage Sludge", Fresenius Z. Anal. Chem. 318 (1984), 588–591.

[322] B. Griepink, H. Muntau, H. Gonska, E. Colinet: "Certification of the Contents of Some Heavy Metals (Cd, Cu, Hg, Ni, Pb and Zn) in Three Types of Soil", Fresenius Z. Anal. Chem. 318 (1984), 490–494.

[323] P. Schramel: "Determination of some additional trace elements in certified standard reference materials (soils, sludges, sediment) by ICP-emission spectrometry", Fresenius Z. Anal. Chem. 333 (1989), 203–210.

[324] M. P. Goudzwaard, M. T. C. de Loos-Vollebregt: "Characterization of noise in inductively coupled plasma – atomic emission spectrometry", Spectrochim. Acta 45B (1990), 887–901.

[325] U. Kurfürst: "Traktat über die „Treffsicherheit" bei der Element-Spurenbestimmung, Teil 1· Bedeutung der Präzision am Beispiel der direkten Feststoffanalyse", GIT Fachz. Lab. 43 (1999), 953–955.

[326]   J. Nölte, F. Scheffler, S. Mann, M. Paul: "The Influence of True Simultaneous Internal Standardization and Background Correction on Repeatability for Laser Ablation and Slurry Technique Coupled to ICP Emission Spectrometry", J. Anal. At. Spectrom. 14 (1999), 597–602.

[327]   Discussion in the Plasmachem-Listserver in May 1999 with contributions from B. Towner, D. Boomer, R. J. Glen, J. Lizane Pamer, G. A. Jenner, B. Brattin, R. A. Peters, R. Stux and E. McCurdy.

[328]   Arbeitsausschuß Chemische Terminologie (AchT) im DIN Deutsches Institut für Normung e. V.: "DIN 32645 – Chemische Analytik; Nachweis-, Erfassungs- und Bestimmungsgrenze; Ermittlung unter Wiederholbedingungen; Begriffe, Verfahren, Auswertung", Beuth Verlag, Berlin 1994.

[329]   L. Brüggemann: "Berechnung ausgewählter Qualitätsmerkmale von Analysenverfahren – Schätzung von Nachweisgrenze und Bestimmungsgrenze", GIT GIT Fachz. Lab. 42 (1998), 696–700.

[330]   G. Lieck: "Nachweisgrenze und Rauschen", LaborPraxis (1998), 62–67.

[331]   P. W. J. M. Boumans, J. J. A. M. Vrakking: "Detection limits in inductively coupled plasma atomic emission spectrometry: an approach to the breakdown of the ratios of detection limits reported for different instruments", Spectrochim. Acta 40B (1985), 1437–1446.

[332]   P. W. J. M. Boumans, R. J. McKenna, M. Bosveld: "Analysis of the limiting noise and identification of some factors that dictate the detection limits in a low-power inductively coupled plasma system", Spectrochim. Acta 36B (1981), 1031–1058.

[333]   T. D. Martin, C. A. Brockhoff, J. T. Creed, EMMC Methods Work Group: "Method 200.7 Revision 4.4 – Determination of metals and trace metals in waters and wastes by inductively coupled plasma-atomic emission spectrometry", U. S. Environmental Protection Agency, Cincinnati, Ohio, USA, 1994.

[334]   W. Funk, V. Dammann, C. Vonderheid, G. Oehlmann (Ed.): "Statistische Methoden in der Wasseranalytik", VCH, Weinheim 1985.

[335]   R. O. Scott, R. Jenkins, L. R. P. Butler, A. M. Ure: "Nomenclature, Symbols, Units and Their Usage in Spectrochemical Analysis – X. Preparation of Materials for Analytical Spectroscopy and Other Related Techniques", Pure and Appl. Chem.60 (1988), 1461-1472.

[336]   S. Knobloch, H. W. Zwanziger: "Probleme und Artefakte bei der chemometrischen Aufbereitung von Umweltdaten", GIT Fachz. Lab. 39 (1995), 535–541.

[337]   K. Danzer: "Probenahme inhomogener Materialien – Teil 1", GIT Fachz. Lab. 39 (1995), 928–938.

[338]   K. Danzer: "Probenahme inhomogener Materialien – Teil 2", GIT Fachz. Lab. 39 (1995), 1019–1023.

[339]   J. N. Miller: "Calibration Methods in Spectroscopy: I. Why Are Calibration Methods Useful in Spectroscopy?", Spectrosc. Int. 3 {3} (1991), 41–43.

[340]   H. Puchelt, T. Nöltner: "Zur Stabiltät hochverdünnter Multielement-Eichstandardlösungen – Erfahrungen mit ICP-MS", Fresenius Z. Anal. Chem. (1988), 216–219.

[341]   D. A. Bass, L. B. TenKate: "Stability of Low-Concentration Calibration Standards for Graphite Furnace Atomic Absorption Spectroscopy", At. Spectrosc. 18 (1997), 1–12.

[342]   J. Dahmen, K. Englert, G. Giebenhain: "Properties and production of labware from fluorinated hydrocarbons and their advantages for ultratrace analysis", Lecture "4th Russian-German-Ukrainian Analytical Symposium, Sofrino, Moscow district, Feb. 25 – March 3, 1996", reprint of VIT-LAB, Seeheim, Germany.

[343]   J. N. Miller: "Calibration Methods in Spectroscopy: II. Is it a Straight Line?", Spectrosc. Int. 3 {4} (1991), 42–44.

[344] J. N. Miller: "Calibration Methods in Spectroscopy: III. Straight-Line Graphs – Assumptions and Equations", Spectrosc. Int. 3 {5} (1991), 43–46.

[345] J. N. Miller: "Calibration Methods in Spectroscopy: IV. Errors in Calibration Graphs", Spectrosc. Int. 3 {7} (1991), 45–47.

[346] J. N. Miller: "Calibration Methods in Spectroscopy: V. Further Errors in Calibration Graphs", Spectrosc. Int. 4 {1} (1992), 41–43.

[347] V. Sixta: "Evaluation of ICP Spectrometric Measurements", At. Spectrosc. 12 (1991), 11–15.

[348] J. Nölte: "Standardadditions-Kalibration in der ICP-AES", Angewandte Atomspektrometrie Nr. 2.10, Überlingen 1991.

[349] M. W. Tikannen, G. Wilson: "Statistical Process Control: A Versatile Tool of Scientific Management", Spectrosc. Int. 4 {7} (1992).

[350] W. Funk, V. Dammann, G. Donnevert: "Qualitätssicherung in der Analytischen Chemie", VCH, Weinheim 1992.

[351] K. Doerffel: "Anwendung elementarer Zeitreihenmodelle bei der Beurteilung von Meß-serien", GIT Fachz. Lab. 40 (1996), 532–534.

[352] U. Steingruber, M. Kaindl: "LIMS im Bayrischen LKA", GIT Fachz. Lab. 45 (2001), 190.

[353] M. Carré, E. Poussel, J.-M. Mermet: "Drift Diagnostics in Inductively Coupled Plasma Atomic Emission Spectrometry", J. Anal. At. Spectrom. 7 (1992), 791–797.

[354] E. Poussel, J. M. Mermet: "Simple experiments for the control, the evaluation and the diagnostics of in inductively coupled plasma sequential systems", Spectrochim. Acta 48B (1993), 743–755.

[355] C. Sartoros, E. D. Salin: "Inductively coupled plasma-atomic emission spectrometer warning diagnosis procedure using blank solution data", Spectrochim. Acta 53B (1998), 741–750.

[356] J. Nölte: "Analyse von alkalischen Lösungen mit der ICP-AES", Angewandte Atomspektrometrie Nr. 2.5, Überlingen, Germany 1989.

[357] J. P. Rybarczyk, C. P. Jester, D. A. Yates, S. R. Koirtyohann: "Spatial Profiles of Inter-element Effects in the Inductively Coupled Plasma", Anal. Chem. 54 (1982), 2162–2170.

[358] M. W. Blades, G. Horlick: "Interference from easily ionizable element matrices in induc-tively coupled plasma emission spectrometry – a spatial study", Spectrochim. Acta 36B (1981), 881–900.

[359] W. Schulz, L. Kotz: "Systematische Fehler bei der extremen Elementspurenanalyse", CLB Chemie für Labor und Betrieb 33 (1982), 483–493.

[360] H. Bendlin: "Reinstwasser im Labor", LABO {5} (1992), 20–26.

[361] H. Träger: "Reinstwasser-Herstellung für das analytische Labor", LABO {4} (1993), 46–53.

[362] T. A. Anderson, A. R. Forster, M. L. Parsons: "ICP Emission Spectra: II. alkaline Earth Elements", Appl. Spectrosc. 36 (1982), 504–509.

[363] S. P. Ericson: "Tips for Aluminum Determination at Ultratrace Levels", At. Spectrosc. 13 (1992), 208–212.

[364] J. Nölte: "Die Bestimmung ausgewählter Elementgehalte in den Korngrößenfraktionen von Flugaschen einer Müllverbrennungsanlage", Diplomarbeit Universität Hamburg 1981.

[365] R. Schelenz, E. Zeiller: "Influence of digestion methods on the determination of total Al in food samples by ICP-ES", Fresenius J. Anal. Chem. 345 (1993), 68–71.

[366] I. B. Brenner, M. M. Cheatham: "Determination of boron: aspects of contamination and vaporization", ICP Inf. Newsl. 24 (1998), 320-321.

[367] T. Ishikawa, E. Nakamura: "Suppression of boron volatilization from a hydrofluoric acid solution using a boron-mannitol complex", Anal. Chem. 62 (1990), 2612–2616.

[368] A. Al-Ammar, R. K. Gupta, R. M. Barnes: "Elimination of born memory effects in inductively coupled plasma-mass spectrometry by addition of ammonia", Spectrochim. Acta 54B (1999), 1077–1084.

[369] R. Wennrich, A. Mroczek, K. Dittrich, G. Werner: "Determination of nonmetals using ICP-AES-techniques", Fresenius J. Anal. Chem. 352 (1995), 461–469.

[370] K. Lewin, J. N. Walsh, D. L. Miles: "Determination of dissolved sulphide in groundwaters by inductively coupled plasma atomic emission spectrometry", J. Anal. At. Spectrom. 2 (1987), 249–250.

[371] A. P. Krushevska, R. M. Barnes: "Determination of Low Silicon Concentrations in Food and Coral Soil by Inductively Coupled Plasma Atomic Emission Spectrometry", J. Anal. At. Spectrom. 9 (1994), 981–984.

[372] P. Heitland: "Application of prominent spectral lines for ICP-OES in the 125–190 nm range", GIT Fachz. Lab. 44 (2000), 847–849.

[373] N. Wieberneit, P. Heitland: "Application of ICP-OES with a New argon-Filled CCD Spectrometer Using Spectral Lines in the Vacuum Ultraviolet Spectral Range", Appl. Spectrosc. 55 (2001), 598–603.

[374] J. Nölte: "Pollution Source Analysis of River Water and Sewage Sludge", Environm. Technol. Lett. 9 (1988), 857–868.

[375] M. Volkmer, J. Nölte: "Umweltanalytik – Klassifizierung und Charakterisierung von Abfällen und Altlasten mit der ICP-Atomemissionsspektrometrie in Anlehnung an DIN-Verfahren", LABO 23 {3} (1992), 22–31.

[376] Amtsblatt der Europäischen Gemeinschaften "Richtlinie 98/83/EG des Rates vom 3. November 1998 über die Qualität von Wasser für den menschlichen Gebrauch".

[377] Tran T. Nham: "Analysis of potable water for trace elements by ICP-AES", ICP-AES Instruments at work, Varian Application ICP-1, May 1991.

[378] H.-J. Hoffmann: "Der Einsatz der simultanen ICP-AES in der Wasseranalytik", LaborPraxis {July/Aug.} (1980), 18–25.

[379] B. Raue, H. J. Brauch, J. Nölte: "Wasseranalytik - Multielementbestimmung in Trink- und Grundwässern mit der sequentiellen ICP-AES", LABO 21 (1990), 78–82.

[380] C.E. Taylor, T. L. Floyd: "Inductively Coupled Plasma-Atomic Emission Spectrometric Analysis of Environmental Samples Using Ultrasonic Nebulization", Appl. Spectrosc. 35 (1981), 408–413.

[381] D. D. Nygaard, F. Bulman: "Analysis of Water for Arsenic, Lead, Selenium, and Thallium by Inductively Coupled Plasma Atomic Emission Spectrometry at Contract Laboratory Program Levels", Spectrosc. Int. 2 {2} (1990), 44–47.

[382] Tran T. Nham: "Water analysis using ICP-AES with an ultrasonic nebulizer", ICP-AES Instruments at work, Varian Application ICP-8, Dec. 1992.

[383] J. Nölte: "ICP-OES-Analyse mit Ultraschallzerstäuber – Toxikologisch relevante Elemente im Trinkwasser", LaborPraxis 17 (1993), 46–50.

[384] B. Raue, H.-J. Brauch, F. H. Frimmel: "Determination of sulfate in natural waters by ICP/OES – comparative studies with ion chromatography", Fresenius J. Anal. Chem. 340 (1991), 395–398.

[385] W. Jäger: "Praktische Erfahrungen mit der Plasma-Emissions-Technik (ICP) in der Routineuntersuchung von Abwasser-, Schlamm- und Bodenproben", Z. Wasser-Abwasser-Forsch. 16 (1983), 231–233.

[386] J. Nölte, W. Schrader: "Analytik von Abwasser und Klärschlamm nach DIN 38406 Teil 22", in: W. Günther, J. P. Matthes (Ed.): "InCom '89", GIT Verlag 1989, 84–94.

[387] J. Nölte: "ICP-OES Analysis of Wastewaters and Sludges", At. Spectrosc. 15 (1994), 223–228.

[388] H.-J. Hoffmann: "Die neue Klärschlammverordnung", LaborPraxis (Aug. 1992), 770–773.

[389] J. Nölte: "Beitrag zur Aufklärung von Verunreinigungen durch verschiedene Schadelemente in Gewässern und Klärschlamm mit Hilfe der optischen Emissionsspektrometrie", Dissertation, Universität Hamburg 1985.

[390] H.-J. Hoffmann: "Spektrometrische Verfahren im Rahmen der rechtlichen Wasseranalytik", LaborPraxis {Aug.} (1992), 790–795.

[391] W. Schrader, H. Hein: "ICP-AES-Analyse für Klärschlamm und Böden", LaborPraxis {Jan./Feb.} (1983), 34–53.

[392] M. M. Mosely, P. N. Vijan: "Simultaneous determination of trace metals in sewage and sewage effluents by inductively coupled argon plasma atomic emission spectrometry", Anal. Chim. Acta 130 (1981), 157–166. + Erratum ibid. 131 (1981), 327.

[393] P. Schramel, X. Li-Qiang, A. Wolf, S. Hasse: "ICP-Emissionsspektroskopie: Ein analytisches Verfahren zur Klärschlamm- und Bodenüberwachung in der Routine", Fresenius Z. Anal. Chem. 313 (1982), 213–216.

[394] H. Jäger, K. Slickers: "Simultane spektrometrische Analyse von Böden und Klärschlämmen mit Spectroflame-ICP", Spectro Applikationsbericht 55, Kleve, Germany.

[395] A. F. Ward, L. F. Marciello, L. Carrara, V. J. Luciano: "Simultaneous Determination of Major, Minor, and Trace Elements in Agricultural and Biological Samples by Inductively Coupled argon Plasma Spectrometry", Spectrosc. Lett. 13 (1980), 803–831.

[396] P. Schramel: "Application of High Resolution ICP-Spectrometry to the Determination of B, Be, Co, Mo and Sn in Soils and Similar Matrices", Mikrochim. Acta III (1989), 355–364.

[397] H. Berger: "Atomspektrometrische Analytik bei der Erkundung kontaminierter Standorte", in: B. Welz (Ed.): „5. Colloquium Atomspektrometrische Spurenanalytik", Überlingen 1989, 675–686.

[398] G. Zeibig, J. Luck: "Umweltanalytik an kontaminierten Böden eines ehemaligen Kupferraffineriegeländes mittels ICP-AES", in B. Welz (Ed.): „5. Colloquium Atomspektrometrische Spurenanalytik", Überlingen 1989, 687–695.

[399] J. Nölte: "Analyse von Bodenproben mit der ICP-AES", LaborPraxis Special (1990), 81–82.

[400] M. Jasinski, G. Schwedt, A. Meyer: "Die ICP-OES für die Vanadium-Speciation in Ackerböden", GIT Fachz. Lab. 41 (1997), 482–485.

[401] M. Leiterer, C. J. Nötzold: "Pflanzennährstoffe und Schwermetalle simultan bestimmen", LaborPraxis {July} (1993), 16–24.

[402] D. Florian, R.M. Barnes, G. Knapp. "Comparison of microwave-assisted acid leaching techniques for the determination of heavy metals in sediments, soils, and sludges", Fresenius J. Anal. Chem. 362 (1998), 558–565.

[403] W. Dannecker: "Anwendung der Atomspektroskopie zur Beurteilung chemischer und ökotoxischer Eigenschaften von Stäuben aus Emissionen und Immissionen", in B. Welz (Ed.): „2. Colloquium Atomspektrometrische Spurenanalytik", Verlag Chemie, Weinheim 1982, 187–211.

[404] K. Naumann: "Differenzierte Probenahme und Analytik von Aerosolen unter Anwendung atomspektrometrischer Methoden – Ein Beitrag für künftige Immissionsüberwachungen", Dissertation Universität Hamburg 1983.

[405] T. C. Thomas, L. J. Jehl: "Metal exposure evaluation: A rapid multielement analysis technique using ICP-AES", At. Spectrosc. 9 (1988), 154–156.

[406] M. Kriews, K. Naumann, W. Dannecker: "Einsatz atomspektrometrischer Methoden zur Multielementbestimmung in marinen Aerosolen", in B. Welz (Ed.): „5. Colloquium Atomspektrometrische Spurenanalytik", Überlingen 1989, 633–646.

[407] F. Meyberg: "Anwendungen der Atomemissionsspektrometrie mit induktiv gekoppeltem Plasma zur Elementbestimmung in umweltrelevanten Proben", Dissertation Universität Hamburg 1986.

[408] M. Kriews, K. Naumann, W. Dannecker: "Immissionsstäube auf Filtermaterial", J. Aerosol. Sc. 19 (1988), 1295–1298.

[409] E. A. Stadlbauer: "Aufschluß der Analysenprobe: Klar und sicher", GIT Fachz. Lab. 40 (1996), 722–723.

[410] Tran T. Nham: "Analysis of coal fly ash by inductively coupled plasma-emission spectrometry", ICP-AES Instruments at work, Varian Application ICP-5, Sept. 1991.

[411] W. Dannecker, U. Düwel: "Schadstoffbilanzierungen an Müllverbrennungsanlagen mit Abgasreinigungen", Schriftenreihe "Angewandte Analytik" herausgegeben von W. Dannecker, Inst. f. Anorg. und Angew. Chemie der Universität Hamburg, Band 1 bis 3, 1986.

[412] J. Nölte: „Analyse von Kohleproben und Flugasche eines Kohlekraftwerkes", Angewandte Atomspektrometrie Nr. 4.6, Überlingen 1989.

[413] M. Betinelli, U. Baroni: "Determination of major and trace elements in copper plant fly ash by ICP emission spectrometry", At. Spectrosc. 9 (1988), 157–165.

[414] F. Meyberg, P. Krause, W. Dannecker: "Luftstaubuntersuchungen unter Berücksichtigung der TA Luft mittels ICP-AES, ICP-MS und GF-AAS", in B. Welz (Ed.): „6. Colloquium Atomspektrometrische Spurenanalytik", Überlingen 1991, 707–723.

[415] Anonymous: "Technische Anleitung zur Reinhaltung der Luft – TA Luft - Erste Allgemeine Verwaltungsvorschrift zum Bundes-Immissionsschutzgesetz vom 15.3.1974".

[416] P. Schramel, X. Li-giang: "Investigation of Interelement Effects in Trace Element Analysis in Biological and Environmental Samples by ICP Spectrometry", ICP Information Newslett. 7 (1982), 429–440.

[417] G. Steffes, J. Luck: "Analyse biologischer Standard-Referenzmaterialien mit dem Spectroflame-ICP"; Spectro Report Applikationsbericht 74, Kleve.

[418] L. Reichelt, I. Kluge: "Pflanzenanalytik mittels IPC", GIT Fachz. Lab. 38 (1994), 778–779.

[419] K. A. Anderson: Micro-Digestion and ICP-AES Analysis for the Determination of Macro and Micro Elements in Plant Tissues", At. Spectrosc. 17 (1996), 30–33.

[420] H. Matusiewicz, R. M. Barnes: "Tree Ring Wood Analysis After Hydrogen Peroxide Pressure Decomposition With Inductively Coupled Plasma Atomic Emission Spectrometry and Electrothermal Vaporization", Anal. Chem. 57 (1985) 406–411.

[421] S. R. Koirtyohann, D. A. Yates: "Determination of Major, Minor, and Trace Elements in NIST Biological Reference Materials", At. Spectrosc. 15 (1994), 167–168.

[422] A. Krushevska, R. M. Barnes, C. Amarasiriwaradena: "Decomposition of Biological Samples for Inductively Coupled Plasma Atomic Emission Spectrometry Using an Open Focused Microwave Digestion System", Analyst 118 (1993), 1175–1181.

[423] T. Liese: "Zur Bestimmung von Elementen in Pflanzen- und Bodenproben mittels sequentieller ICP-AES", Fresenius Z. Anal. Chem. 321 (1985), 37–44.

[424] B. Madeddu, A. Rivoldini: "Analysis of Plant Tissues by ICP-OES and ICP-MS Using an Improved Microwave Oven Acid Digestion", At. Spectrosc. 17 (1996), 148–154.

[425] F. J. Copa-Rodríguez, M. I. Basadre-Pampín: "Determination of iron, copper and zinc in tinned mussels by inductively coupled plasma atomic emission spectrometry (ICP-AES)", Fresenius J. Anal. Chem. 348 (1994), 390–395.

[426] H. Schorin, Z. Benzo, E. Marcano, C. Gomez, F. O. Bamiro: "Accurate and Precise Trace Element Determination in Biomonitors Using ICP-OES"; At. Spectrosc. 19 (1998), 129–132.

[427] J. Locke: "The application of plasma source atomic emission spectrometry in forensic science", Anal. Chim. Acta 113 (1980), 3–12.

[428] P. Schramel, B.-J. Klose: "Direct determination of Cu, Fe, Zn, Ca, Mg and Na in serum by means of ICP - emission spectrometry", Fresenius Z. Anal. Chem. 307 (1981), 26-30.

[429] P. Schramel: "Consideration of inductively coupled plasma spectroscopy for trace element analysis in the bio-medical and environmental fields", Spectrochim. Acta 38B (1983), 199-206

[430] H. S. Mahanti, R. M. Barnes: "Determination of major, minor and trace elements in bone by inductively-coupled plasma emission spectrometry", Anal. Chim. Acta 151 (1983), 409-417.

[431] G. Morisi, M. Patriarca, F. Petrucci, L. Fornarelli, S. Caroli: "Reliability of Inductively Coupled Plasma - Atomic Emission Spectrometry Determinations of Urinary Electrolytes Compared with Flame Atomic Absorption Spectrometry", Spectroscopy Int. 2 {5} (1990), 32–38.

[432] G. Crisponi, V. M. Nurchi, R. Silvagna, G. Lubinu, R. Ambu, A. Marras, G. Parodo, G. Faa: "Critical Evaluation of Analytical Procedures for Trace-Element Determination in Human Liver Using ICP-OES", At. Spectrosc. 16 (1995), 73–78.

[433] C. Prohaska, K. Pomazal, I. Steffan: "Determination of Ca, Mg, Fe, Cu, and Zn in blood fractions and whole blood of humans by ICP-OES", Fresenius J. Anal. Chem. 367 (2000), 479-484.

[434] R. Giordano, S. Costantini, I. Vernillo, B. Casetta, F. Aldrighetti: "Comparative study for aluminium determination in bone by atomic absorption techniques and inductively coupled plasma atomic emission spectroscopy", Microchem. J. 30 (1984), 435–447.

[435] F. E. Lichte, S. Hopper, T. W. Osborn: "Determination of Silicon and Aluminum in Biological Matrices by Inductively Coupled Plasma Emission Spectrometry", Anal. Chem. 52 (1980), 120–124.

[436] P. Schramel, X. Li-Quang: "Determination of Beryllium in the Parts-per-Billion Range in Three Standard Reference Materials by Inductively Coupled Plasma Atomic Emission Spectrometry", Anal. Chem. 54 (1982), 1333–1336.

[437] A. Sanz-Medel, R. Rodriguez Roza, C. Perez-Conde: "A Critical Comparative Study of Atomic-spectrometric Methods (Atomic Absorption, Atomic Emission and Inductively Coupled Plasma Emission) for Determining Strontium in Biological Materials", Analyst 108 (1983), 204–212.

[438] Kunio Shiraishi, Hisao Kawamura, Gi-Ichiro Tanaka: "Determination of alkaline-earth metals in foetus bones by inductively-coupled plasma atomic emission spectrometry", Talanta 34 (1987), 823–827.

[439] J. D. Kruse-Jarres: "Grenzen der Spurenelement-Analytik", GIT Labor-Medizin {1} (1995), 67–71.

[440] P. Schramel, B.-J. Klose: "Direktbestimmung von Cu, Fe, Zn, Ca, Mg und Na im Serum mittels ICP-Emissionsspektralanalyse", Fresenius Z. Anal. Chem. 307 (1981), 26–30.

[441] D. E. Nixon, T. P. Moyer, P. Johnson, J. T. McCall, A. B. Ness, W. H. Fjerstad, M. B. Wehde: "Routine Measurement of Calcium, Magnesium, Copper, Zinc, and Iron in Urin and Serum by Inductively Coupled Plasma Emission Spectroscopy", Clin. Chem. 32 (1986), 1660–1665.

[442] R. F. M. Herber, H. J. Pieters, J. W. Elgersma: "A Comparison of Inductively Coupled argon Plasma Atomic Emission Spectrometry and Electrothermal Atomization Atomic Absorption Spectrometry in the Determination of Copper and Zinc in Serum", Fresenius Z. Anal. Chem. 313 (1982), 103–107.

[443] Mian-zhi Zhuang, R. M. Barnes: "Determination of Major, Minor, and Trace Elements in Human Serum by Using Inductively Coupled Plasma-Atomic Emission Spectroscopy", Appl. Spectrosc. 39 (1985), 793–796

[444] C. Minoia, L. Pozzoli, S. Angeleri, G. Tempini, F. Candura: "Determination of Silicon in urine by inductively coupled plasma emission spectroscopy", At. Spectrosc. 3 (1982), 70–72.

[445] T. Tanaka, Y. Hayashi: "Determination of silicon, calcium, magnesium and phosphorus in urine using inductively coupled plasma emission spectrometry and a matrix-matching technique", Clin. Chim. Acta 156 (1986), 109–113.

[446] G. Morisi, M. Patriarca, F. Petrucci, L. Fornarelli, S. Caroli: "The Reliability of Inductively Coupled Plasma-Atomic Emission Spectrometry Determination of Urinary Electrolytes Compared with Flame Atomic Absorption Spectrometry", Spectrosc. Int. 2 {5} (1990), 32–38.

[447] M. M. Kimberly, D. C. Paschal: "Screening for selected toxic elements in urine by sequential-scanning inductively-coupled plasma atomic emission spectrometry", Anal. Chim. Acta 174 (1985), 203–210.

[448] M. López-Artíguez, A. Cameán, M. Repetto: "Preconcentration of Heavy Metals in Urine Using Chelating Ion Exchange Resin and Quantification by ICP-AES", At. Spectrosc. 17 (1996), 83–87.

[449] R. W. Kuennen, K. A. Wolnik, F. L. Fricke, J. A. Caruso: "Pressure Dissolution and Real Sample Matrix Calibration for the Multielement Analysis of Raw Agricultural Crops by Inductively Coupled Plasma Atomic Emission Spectrometry", Anal. Chem. 54 (1982), 2146–2150.

[450] K. W. Barnes: "Determinination of Nutrition Labeling Education Act Minerals in Foods by ICP-OES", At. Spectrosc. 18 (1997), 41–54.

[451] P. Schramel: "Bestimmung von Bor in Milch und Milchprodukten mittels ICP-Emissionsspektralanalyse", Z. Lebensm. Unters. Forsch. 169 (1979), 255–258.

[452] A. Krushevska, R.M. Barnes, C.J. Amarasiriwaradena, H. Foner, L. Martines: "Comparison of Sample Decomposition Procedures for the Determination of Zinc in Milk by Inductively Coupled Plasma Atomic Emission Spectrometry", J. Anal. At. Spectrom. 7 (1992), 851-858.

[453] K. W. Barnes: "Determining Nutrients in Dairy Products with ICP-OES", Food Quality {June} (1996).

[454] M. Thomsen, R. Vernoy, M. Volkmer: "Analyse von Milchpulver mittels ICP-Emissionsspektrometrie", GIT Fachz. Lab. 42 (1998), 894–895.

[455] S. Nikdel, R. D. Carter: "Determination of the Mineral content of orange juice using automated fast sequential ICP-AES", International Laboratory {Oct.} (1986), 32–42.

[456] K. W. Barnes: "Trace Metal Determinations in Fruit, Juice, and Juice Products Using an Axially Viewed Plasma", At. Spectrosc. 18 (1997), 84–101.

[457] H. Eschnauer, L. Jakob, H. Meierer, R. Neeb: "Use and Limitations of ICP-OES in Wine Analysis", Mikrochim. Acta [Wien] III (1989), 291–298.

[458] W. M. Blakemore, S. M. Billedeau: "Analysis of Laboratory Animal Feed for Toxic and Essential Elements by Atomic Absorption and Inductively Coupled argon Plasma Emission Spectrometry", J. Assoc. Off. Anal. Chem. 64 (1981), 1284–1290.

[459] P. A. Liberatore: "Determination of trace elements in geological samples by ICP-AES", ICP-AES Instruments at work, Varian Application ICP-16, Sept. 1994.

[460] P. A. Liberatore: "Determination of majors in geological samples by ICP-AES", ICP-AES Instruments at work, Varian Application ICP-12, July 1993.

[461] H. Heinrichs, A. G. Herrmann: "Praktikum der Analytischen Geochemie", Springer, Berlin Heidelberg 1990.

[462] J. Broekaert: "Der Einsatz des ICP für die Analyse von geochemischen Proben", LaborPraxis (Jan./Feb. 1980), 43–49.

[463] G. F. Wallace: "Application of a sequential scanning ICP to the analysis of geological materials", At. Spectrosc. 2 (1981), 87–90.

[464] R. M. Barnes, H. S. Mahanti: "Analysis of bauxite by inductively coupled plasma-atomic emission spectroscopy", Spectrochim. Acta 38B (1983), 193–197.

[465] N. Korte, M. Kollenbach, S. Donivan: "The determination of Uranium, Thorium, Yttrium, Zirconium and Hafnium in Zircon", Anal. Chim. Acta 146 (1983), 267–270.

[466] B. Casetta, A. Giaretta, G. Rampazzo: "An approach to ICP analysis of geological samples", At. Spectrosc. 4 (1983), 152–154.

[467] Isaac B. Brenner, S. Erlich: "The Spectrochemical (ICP-AES) Determination of Tungsten in Tungsten Ores, Concentrates and Alloys: An Evaluation as an Alternative to the Classical Gravimetric Procedure", Appl. Spectrosc. 38 (1984), 887–890.

[468] A. M. Marabini, M. Barbaro, B. Passariello: "Determination of Uranium in rocks and minerals by plasma emission spectroscopy", At. Spectrosc. 6 (1985), 74–75.

[469] R. Saran, N. K. Baishy: "Direct Determination of Thorium in Geological Samples Using ICP-OES", At. Spectrosc. 18 (1997), 60–63.

[470] J. A. C. Broekaert, F. Leis, K. Laqua: "Application of an inductively coupled plasma to the emission spectoscopic determination of rare earths in mineralogical samples", Spectrochim. Acta 34B (1979), 73–84.

[471] I. B. Brenner, A. E. Watson, T. W. Steele, A. A. Jones, M. Goncalves: "Application of an argon-nitrogen inductively-coupled radiofrequency plasma (ICP) to the analysis of geological and related materials for their rare earth contents", Spectrochim. Acta 36B (1981), 785–797.

[472] J. G. Crock, F. E. Lichte: "Determination of the REE and Y in silicate materials by Inductively Coupled argon Plasma/Atomic Emission Spectrometry", Anal. Chem. 54 (1982), 1329–1332.

[473] E. Zuleger, J. Erzinger: "Determination of the Rare Earth Elements in Geological Materials by ICP-AES", Fresenius Z. Anal. Chem. 332 (1988), 140–143.

[474] Z. Sulcek, I. Rubeska, V. Sixta, T. Paukert: "Determination of rare earth elements and Yttrium in rocks using the Plasma II ICP emission spectrometer", At. Spectrosc. 10 (1989), 4–9.

[475] I. Roelandts: "Determination of Ten Rare Earth Elements in Rare Earth Ores by Inductively Coupled Plasma – Atomic Emission Spectrometry", At. Spectrosc. 13 (1992), 193–198.

[476] A. K. Das, P. Roychowdhury: "Determination of Rare Earth Elements in Bauxite by ICP-AES", At. Spectrosc. 18 (1997), 80–83.

[477] A. F. Ward, L. F. Marciello: "Analysis of Metal Alloys by Inductively Coupled argon Plasma Optical Emission Spectrometry", Anal. Chem. 51 (1979), 2264–2272.

[478] W. Blödorn, T. Rippert, D. Thierig, H. Unger: "Spuren- und Elementanalyse von Stählen und Ferrolegierungen – Emissionsspektrometrie mit ICP-Anregung – Teil 2", LaborPraxis (Okt. 1992), 982–987.

[479] W. Blödorn, T. Rippert, D. Thierig, H. Unger: "Spuren- und Elementanalyse von Stählen und Ferrolegierungen – Emissionsspektrometrie mit ICP-Anregung – Teil 1", LaborPraxis (Sept. 1992), 878–886.

[480] I. Hlavácek, I. Hlaváckova: "Determination of Boron in Boron-Alloyed Steels by Inductively Coupled Plasma Atomic Emission Spectrometry", Mikrochim. Acta [Wien] III (1989), 309–314.

[481] I. Novozamsky, R. van Eck, J. J. van der Lee, V. J. G. Houba, G. O. Ayaga: "Continuous-flow technique for generation and separation of Methyl Borate from Iron-containing

matrices with subsequent determination of Boron by ICP-AES", At. Spectrosc. 9 (1988), 97–99.

[482] I. Novozamsky, R. van Eck, V. J. G. Houba, J. J. van der Lee: "A New Solvent Extraction for the Determination of Traces of Boron by ICP-AES", At. Spectrosc. 11 (1990), 83–84.

[483] J. Ciba, B. Smolec: "Determination of boron in steel by an ICP emission spectrometric technique", Fresenius J. Anal. Chem. 348 (1994), 215–217.

[484] J. Nölte: "Anwendung chemometrischer Rechentechniken zur Korrektur spektraler Störungen in einer Eisen-Matrix", in: VDEh (Ed.): Tagungsband 21. Spektrometertagung '96, 1997, 193–206.

[485] D. E. Gillum, H. P. Vail: "Wavelength Selection for the Analysis of Ferrous Materials by Inductively Coupled Plasma Spectrometry", Perkin Elmer, Norwalk, USA, 1989.

[486] P. S. Doidge: "Determination of trace impurities in high-purity copper by sequential ICP-AES with axial viewing", ICP-AES Instruments at work, Varian Application ICP-25, November 1998.

[487] A. D. King, G. F. Wallace: "Determination of trace elements in copper metal, copper oxides, and bronze by ICP emission spectrometry", At. Spectrosc. 6 (1985), 4–8.

[488] I. Segal, A. Kloner, I. B. Brenner: "Multi-element Analysis of Archaeological Bronze Objects Using Inductively Coupled Plasma Atomic Emission Spectrometry: Aspects of Sample Preparation and Spectral Line Selection", J. Anal. At. Spectrom. 9 (1994), 737–742.

[489] I. Steffan, G. Vujicic: "Analysis of Zirconium Alloys by Inductively Coupled Plasma Atomic Emission Spectrometry", J. Anal. At. Spectrom. 9 (1994), 785–789.

[490] Shang-Jing Lu, R. M. Barnes: "Determination of Cadmium in Pure Zirkonium by Inductively Coupled Plasma - Atomic Emission Spectroscopy", Appl. Spectrosc. 38(1984) 284–285.

[491] J. Nölte, H. Kima, A. Meyer, W. Wolff: "Präzisionsbestimmungen von Edelmetallen auf Autokatalystoren mit einem Array-ICP-Emissionsspektrometer", GIT Fachz. Lab. 44 (2000), 839–841.

[492] M. Brill, K.-H. Wiedemann: "Determination of Gold and Gold Alloys by ICP Spectrometry", Gold Bull. 25 {1} (1992), 13–26.

[493] International Organization for Standardization: "Determination of Gold in Jewellery White Gold Alloys – ICP-Solution-Spectrometric Method", Doc. ISO/TC 174 N 71, 1992.

[494] International Organization for Standardization: "Determination of Platinum in Platinum Jewellery Alloys – ICP-Solution-spectrometric Method using Yttrium as Internal Standard Element", ISO/DIS 11494.2, 1997.

[495] International Organization for Standardization: "Determination of Palladium in Palladium Jewellery Alloys – ICP-Solution-spectrometric Method Using Yttrium as Internal Standard Element", ISO/DIS 11495.2, 1997.

[496] A. Marucco, C. Marcolli, R. Magarini: "Analysis of Gold Alloys by ICP Emission Spectrometry With Use of Yttrium or Indium as Internal Standard," At. Spectrosc. 20 (1999), 134–141.

[497] H.-M. Lüschow: "Gold analysis in a precious metal refinery – state and development", Erzmetall 50 (1997), 22–42.

[498] H.-M. Lüschow: "Edelmetallanalytik – Eine Übersicht über bewährte klassische und moderne Methoden", GDMB Schriftenreihe 69 (1993).

[499] M. Baucells, G. Lacort, M. Roura: "Determination of Trace Amounts of Palladium, Iron and Copper in Pure Gold by Inductively Coupled Plasma Atomic Emission Spectrometry", J. Anal. At. Spectrom. 2 (1987), 645–647.

[500] H.-M. Lüschow: "Purity Analysis of Silver Nitrate", Erzmetall 51 (1998), 195–205.

[501] S. Calmotti, C. Dossi, S. Recchia, G. M. Zanderighi: "Solving the problems in noble metals analysis via inductively coupled plasma atomic emission spectroscopy (ICP/AES)", Spectrosc. Eur. 8 {6} (1996), 18–22.

[502] J. Nölte: "ICP-Emissionsspektrometrie zur Platin-Bestimmung", CLB Chemie in Labor und Biotechnik 51 (2000), 376–380.

[503] N. Korte, M. Hollenbach, S. Donivan: "Determination of Zr:Hf ratios by inductively coupled plasma emission spectroscopy", At. Spectrosc. 3 (1982), 79–80.

[504] I. B. Brenner, S. Erlich, G. Vial, J. McCormack, P. Grosdaillon, A. E. Asher: "Direct Trace Element Analysis of Tungsten Powders, Alloys and Related Materials by Inductively Coupled Plasma Atomic Emission Spectrometry (ICP-AES)", J. Anal. At. Spectrom. 2 (1987), 637–644.

[505] C. W. Whitten: "ICP analysis of recycled superalloy scrap", At. Spectrosc. 8 (1987), 81–83.

[506] M. Renko, A. Osojnik, V. Hudnik: "ICP emission spectrometric analysis of rare earth elements in permanent magnet alloys", Fresenius J. Anal. Chem. 351 (1995), 610–613.

[507] T. Piippanen, J. Jaatinen, J. Tummavuori: "The analysis of chromium, cobalt, iron, nickel, niobium, tantalum, titanium and zinc in cemented tungsten carbides with cobalt as binder by inductively coupled plasma atomic emission spectrometry", Fresenius J. Anal. Chem. 357 (1997), 405–410.

[508] M. Uchiyama, D. Yates: "The Analysis of Fe/Nd Alloys", ICP Application Study # 59, PerkinElmer Corporation, Wilton, Connecticut, USA, 1992.

[509] D. Hilligoss, P. Krampitz, D. A. Yates: "ICP-OES Analysis of Complex Alloys Containing Ni, Cr, Cu, and Al", At. Spectrosc. 15 (1994), 254–256.

[510] Yang Xiuhuan, Wei Jinfang, Liu Hongtao, Tang Baoying, Zhang Zhanxia: "Direct determination of trace elements in tungsten products using an inductively coupled plasma optical emission charge coupled device detector spectrometer", Spectrochim. Acta 53B (1998), 1405–1412.

[511] I. Segal, E. Dorfman, O. Yoffe, I. Bezrukavnikov, A. J. Agranat: "Direct ICP-OES Determination of Low Concentrations of Cu, Fe, Li, Ti, and V in Potassium Lithium Tantalate Niobate Crystals", At. Spectrosc. 21 (2000), 46–49.

[512] W. Blödorn, R. Krismer, H. M. Ortner, J. Stummeyer: "Trace-matrix separations for high-purity chromium, molybdenum and tungsten with cellulose collectors", Mikrochim. Acta [Wien] III (1989), 423–432.

[513] E. A. Fitzgerald, A. A. Bornstein, L. J. Davidowski: "Determination of trace elements in negative photoresist by inductively coupled plasma atomic emission spectroscopy and atomic absorption spectroscopy", At. Spectrosc. 6 (1985), 1–3.

[514] J. C. Farinas, M. Barba: "Chemical Analysis by Inductively Coupled Plasma Atomic Emission Spectrometry of Semiconducting Ceramics of Barium Titanate Doped with Various Metal Oxides", Mikrochim. Acta [Wien] III (1989), 299–308.

[515] E. Grallath, P. Tschöpel, G. Kölbin, U. Stix, G. Tölg: "Zur Spektralphotometrie und Emissionsspektrometrie mit Plasmaanregung (CMP, ICP) von Bor-Spuren in Metallen, Silicium und Quarz nach HF-Aufschluß und Abtrennung durch BF_3-Destillation bzw. Ausschütteln von BF_4^--Ionen-Assoziaten", Fresenius Z. Anal. Chem. 302 (1980), 40–51.

[516] M. Zins: "Technische Keramik in Deutschland", GIT Fachz. Lab. 45 (2001), 536–537.

[517] M. Bettinelli, U. Baroni, F. Bilei, G. Bizzarri: "Characterization of SCR-DENOx Materials by ICP-MS and ICP-AES: Comparison with XRF and NAA", At. Spectrosc. 18 (1997), 71–79.

[518] J. L. Fabec, M. L. Ruschak: "Determination of Ni, V, Fe, Na, Ti, Al, Sb, and Sn in fluid catalytic cracking catalyst by ICP/AES - flame AAS ", At. Spectrosc. 6 (1985), 81–87.

[519] S. Mann, D. Geilenberg, J. A. C. Broekaert, P. Kainrath, D. Weber: "Digestion and Characterization of Ceramic Materials and Noble Metals", At. Spectrosc. 19 (1998), 62–65.

[520] J. B. Jones: "Simultaneous Determination of Total Boron, Calcium, Copper, Iron, Magnesium, Manganese, Phosphorus, Potassium, and Zinc in Fertilizers by Inductively Coupled argon Plasma Emission Spectroscopy", J. Assoc. Off. Anal. Chem. 65 (1982), 781–785.

[521] M. L. Fernandez Sanchez, C. Garcia Ortiz, S. Arribas Jimeno, A. Sans-Medel: "Application of the Inductively Coupled Plasma to the Determination of Lead, Zinc, and Silver in Lead and Zinc Flotation concentrates", At. Spectrosc. 5 (1984), 197–203.

[522] A. Bettero, B. Casetta, F. Galiano, E. Ragazzi, C. A. Benassi: "Rheological and Spectroscopic Behavior of Cosmetic Products", Fresenius Z. Anal. Chem. 318 (1984), 525–527.

[523] B. Fairman, M. W. Hinds, S. M. Nelms, D. M. Penny: "Atomic Spectrometry Update - Industrial analysis: metals, chemicals and advanced materials", J. Anal. At. Spectrom. 13 (1998), 233R–266R.

[524] J. Nölte: "Einsatz der Atomemissionsspektrometrie mit induktiv-gekoppeltem Plasma (ICP-AES) zur Analyse von Galvanikbädern", Galvanotechnik 81 (1990), 2738–2741.

[525] F. Finotelle, P. Liberatore: „Direkte Bestimmung von Verunreinigungen in Salzlösungen", LaborPraxis {May} (1998), 82–84.

[526] B. Casetta, A. Giaretta: "ICP analysis of cement", At. Spectrosc. 6 (1985), 144–148.

[527] K.-H. Ebert: "Analysis of Portland Cement by ICP-AES", At. Spectrosc. 16 (1995), 102–103.

[528] L. Paama, P. Peramäki: "Matrix Effects Due to Calcium in argon Plasma Analysis of Calcitic Mortars by ICP-OES", At. Spectrosc. 18 (1997), 119–121.

[529] S. Koelling, J. Kunze: "Analytik der Zusammensetzung von antiken Gläsern", GIT Fachz. Lab. 38 (1994), 1119–1122.

[530] A. Casoli, P. Mirti: "The analysis of archaeological glass by inductively coupled plasma optical emission spectroscopy", Fresenius J. Anal. Chem. 344 (1992), 104–108.

[531] J. Latino: "The Analysis of NIST Glass Materials", At. Spectrosc. 15 (1994), 143–144.

[532] Europäisches Komitee für Normung: "Europäische Norm EN 71: Sicherheit von Spielzeug; Teil 3: Migration bestimmter Elemente", Brüssel 1988.

[533] J. Nölte: "Bestimmung 'toxischer' Elemente in Spielwaren nach EN 71 Teil 3", in: B. Welz (Ed.): "6. Colloquium Atomspektrometrische Spurenanalytik", Überlingen 1991.

[534] P. S. Murty, R. M. Barnes: "Determination of Trace Rare Earth Elements in Uranium by Inductively Coupled Plasma Atomic Emission Spectrometry", J. Anal. At. Spectrom. 1 (1986), 145–148.

[535] Kaneharu Kato: "Application of inductively coupled plasma atomic emission spectrometry to analysis of radioactive materials: A review", At. Spectrosc. 7 (1986), 129–147.

[536] D. G. Weir, M. W. Blades: "Characteristics of an Inductively Coupled argon Plasma Operating with Organic Aerosols. Part 3. Radial Spatial Profiles of Solvent and Analyte Species", J. Anal. At. Spectrom. 11 (1996), 43–52.

[537] A. W. Boorn, R. F. Browner: "Effects of Organic Solvents in Inductively Coupled Plasma Atomic Emission Spectrometry", Anal. Chem. 54 (1982), 1402–1410.

[538] R. I. Botto: "Matrix interferences in the analysis of organic solutions by inductively coupled plasma-atomic emission spectrometry", Spectrochim. Acta 42B (1987), 181–199.

[539] A. Nobile, R. G. Shuler, J. E. Smith: "A modified inductively coupled plasma torch for use with methanol solvents", At. Spectrosc. 3 (1982), 73–75.

[540] J. Schickli, C. Anderau, D. Yates: „Analysis of wear metals in lubricating oils", Technical Summary, PerkinElmer Corp. Norwalk, Conn., USA 1988.

[541]   J. D. Algeo, D. R. Heine, H. A. Phillips, F. B. G. Hoek, M. R. Schneider, J. M. Freelin, M. B. Denton: "On the direct determination of metals in lubricating oils by ICP", Spectrochim. Acta 40B (1985), 1447–1456.

[542]   J. L. Fischer, C. J. Rademeyer: "Direct Determination of Metals in Oils by Inductively Coupled Plasma Atomic Emission Spectrometry Using High Temperature Nebulization", J. Anal. At. Spectrom. 9 (1994), 623–628.

[543]   G. F. Wallace, R. D. Ediger: "Optimization of ICP operating conditions for the determination of Sulfur in oils", At. Spectrosc. 2 (1981), 169–172.

[544]   D. Sommer, K. Ohls: "Organische Lösungsmittel in der ICP-Spektroskopie", LaborPraxis (June 1982), 598–610.

[545]   C. Anderau, K. J. Fredeen, M. Thomsen, D. A. Yates: "Analysis of Wear Metals in Oil by ICP-OES", At. Spectrosc. 16 (1995), 79–81.

[546]   B. Magyar, P. Lienemann, S. Wunderli: "Reduzierung der Untergrundemission eines induktiv gekoppelten Argonplasmas durch Sauerstoffzufuhr bei der Spektralanalyse organischer Lösungen", GIT Fachz. Lab. 26 (1982), 541–548.

[547]   R. Ender: "Die Analyse von gebrauchten Mineralölprodukten mittels ICP", CLB Chemie in Labor und Biotechnik 44 (1993), 232–234.

[548]   A. D. King, D. R. Hilligoss, G. F. Wallace: "Comparison of results for determination of wear metals in used lubricating oils by flame atomic absorption spectrometry and inductively coupled plasma emission spectrometry", At. Spectrosc. 5 (1984), 189–191.

[549]   D. Marquardt: "Organische Elementspurenanalytik per ICP", LaborPraxis {June} (2000), 90–95.

[550]   M. Thomsen, P. Kainrath: "The Analysis of Coal Tar Pitch by ICP Optical Emission Spectrometry After Digestion in a Microwave Oven System", At. Spectrosc. 19 (1998), 60–61.

[551]   J. Nölte: "Rationeller Einsatz atomspektrometrischer Verfahren in der Elementanalytik", Lecture at the conference "Analytiktreffen", Neubrandenburg, Germany 1990.

[552]   W. Slavin: "A Comparison of Atomic Spectroscopic Analytical Techniques", Spectrosc. Int. 4 {1} (1992), 22–27.

[553]   G. Tyler: "ICP-MS, or ICP-AES and AAS? – A comparison", Spectrosc. Eur. 7 {1} (1995), 14–22.

[554]   J.-M. Mermet, E. Poussel: "ICP Emission Spectrometers: 1995 Analytical Figures of Merit", Appl. Spectrosc. 49 (1995), 12A–18A.

Index

9 783527 306725